MOOD
Management
LEADER'S MANUAL

MOOD
Management
LEADER'S MANUAL
A Cognitive-Behavioral Skills-Building Program
for Adolescents

Carol A. Langelier

Sage Publications
International Educational and Professional Publisher
Thousand Oaks ▪ London ▪ New Delhi

For information:

Sage Publications, Inc.
2455 Teller Road
Thousand Oaks, California 91320
E-mail: order@sagepub.com

Sage Publications Ltd.
6 Bonhill Street
London EC2A 4PU
United Kingdom

Sage Publications India Pvt. Ltd.
M-32 Market
Greater Kailash I
New Delhi 110 048 India

Printed in the United States of America

Library of Congress Cataloging-in-Publication Data

Langelier, Carol. A.
    Mood management leader's manual: A cognitive-behavioral skills building program for adolescents / Carol A. Langelier
      p. cm.
    ISBN 0-7619-2297-0 (paperback)
    1. Emotions in adolescence—Juvenile literature.   I. Title
  BF724.3.E5 L36 2000
  155.5'124—dc21   00-010497

05 06 07 7 6 5 4 3

Acquiring Editor:	Nancy Hale
Production Editor:	Nevair Kabakian
Editorial Assistant:	Candice Crosetti
Cover Designer:	Michelle Lee

Clip Art

All clip art used in this workbook was obtained from the following sources:

- Corel Graphics Pack 1995
  Corel Corporation

- Microsoft Publisher 97 CD Deluxe
  Microsoft Corporation

To my family, friends, and mentors
for their support, wisdom, and
belief in emotional wellness.

Table of Contents

How to Use the Skills Workbook

About Mood Management

- **Description**

Mood Management is a skills-building program designed to help adolescents learn to effectively manage difficult emotions such as anger, depression, anxiety, and low self-esteem. On the basis of cognitive behavioral counseling theory, Mood Management emphasizes the importance of developing useful strategies that empower adolescents to make responsible behavioral choices in answer to the challenging emotions they experience.

Often overwhelmed by life's demands, teenagers frequently lack the skills required to effectively maneuver the emotional challenges of adolescence. Mood Management, aptly described as a journey toward emotional wellness, is a program that requires adolescents to learn and practice a set of skills that equip them with the ability to respond to these challenges and accompanying emotions in a healthy manner. Rather than feeling "controlled by" their emotions, adolescents who learn and practice Mood Management skills are enabled to "take control" to effectively manage difficult emotions and corresponding behaviors.

Mood Management encourages adolescents to take responsibility for their emotional well-being. It stresses the importance of practicing the skills introduced in each unit of the *Skills Workbook* to achieve and maintain emotional wellness. Through the use of cognitive behavioral strategies, adolescents learn to challenge self-defeating thoughts that keep them "stuck" in difficult emotions such as anger, depression, anxiety, and low self-esteem.

- **Classroom Guidance, Group Counseling, and Individual Therapy**

Mood Management can be used in schools as the curriculum in a classroom guidance program or group counseling program. In addition, it can serve as a tool used by clinicians working with adolescents in brief, individual therapy or group therapy in both outpatient and inpatient settings. Although the *Skills Workbook* is primarily written from the perspective of a classroom guidance or group counseling program, clinicians can easily adapt the information in the *Skills Workbook* to facilitate individual therapy.

- **The Setting**

Mood Management can be used in many different settings, including schools, inpatient/outpatient treatment facilities, private practice, and community service agencies. School counselors, teachers, psychologists, social workers, and other mental health practitioners who desire to move adolescents toward emotional wellness by emphasizing cognitive behavioral strategies will find Mood Management to be a valuable tool in their work with adolescents.

- **Length of the Mood Management Program**

The number of sessions required to complete the Mood Management Program will vary depending on many different factors such as time allotted per session, format (classroom guidance, group counseling, and individual counseling/therapy), general discussion during each session, etc. The Mood Management Program can generally be completed in 8 to 12 sessions. Adolescents often request to repeat the program and are encouraged to do so to reinforce the skills they learned when taking the program for the first time.

How to Use the Skills Workbook

Getting Started
- **Design**

Each unit in the *Skills Workbook* is designed to introduce adolescents to various Mood Management techniques using a format that is both didactic and experiential in nature. Beginning with the Introduction on page 1, the *Skills Workbook* provides easy-to-follow guidelines for structuring a classroom guidance program as well as group or individual counseling sessions.

- **The Basics**

The format of each unit in the *Skills Workbook* follows a specific sequence as follows:

- ❑ Reading
- ❑ Discussion Questions
- ❑ Skill Session
- ❑ Weekly Assignment
- ❑ Before We Move On - Review & Preview

Each of these sections is easily recognized in the *Skills Workbook* by a special logo. Beginning with Unit 2, the Before We Move On - Review & Preview section is used as the starting point for introducing each new unit in the *Skills Workbook*.

The Table of Contents serves as a handy outline, set off section by section, for the user to follow.

- **Overhead Transparencies**

The presentation of didactic information is aided through the use of overhead transparencies. The user should refer to the List of Transparencies on page xii to determine when each transparency should be introduced and used in the *Skills Workbook*. The Guide to Transparencies provides a visual reference to each transparency. Transparency masters, found in Appendix D, are printed on perforated paper so the user can remove them to make overhead transparencies.

How to Use the Skills Workbook

Guide to Transparencies

How to Use the Skills Workbook

How to Use the Skills Workbook

List of Transparencies

1.	The 3 Yes Rule	Unit 1 - page 9
2.	It's All About IMAGE	Unit 2 - page 16
3.	Making General Goals More Specific	Unit 2 - page 18
4.	Emotional Traffic Jams: The Five Roads	Unit 3 - page 24
5.	Wellness Mind - Emotional Mind	Unit 4 - page 35
6.	It's the Thought That Counts!	Unit 4 - page 36
7.	Thoughts, Feelings, & Behaviors	Unit 4 - page 38
8.	Negative Emotional Cycles	Unit 5 - page 48
9.	The Cycle of Depression	Unit 5 - page 49
10.	The Cycle of Anger	Unit 5 - page 50
11.	The Cycle of Anxiety	Unit 5 - page 51
12.	The Cycle of Low Self-Esteem	Unit 5 - page 52
13.	Characteristics of the Emotional Mind	Unit 6 - page 70
14.	Challenging: Be Your Own Mood Police	Unit 6 - page 72
15.	Challenging: Take a Detour From Your Emotional Mind	Unit 6 - page 73
16.	Challenging Map: The Route to the Wellness Mind	Unit 6 - page 74
17.	The Thought Record	Unit 7 - page 90
18.	My Action Plan...	Unit 7 - page 93

1.

The "3 Yes" Rule

- ☑ Do my goals involve changing myself rather than expecting others to change?

- ☑ Do my goals involve changing things that are within my control?

- ☑ Are my goals realistic?

Mood Management Skills Workbook - Unit 1 - Page 5

2.

It's All About IMAGE

- ✓ <u>I</u> do care.
- ✓ <u>M</u>anaging problems effectively is a plus.
- ✓ <u>A</u>wareness is important.
- ✓ <u>G</u>o for it - it's worth the effort!
- ✓ <u>E</u>motions - we all have them and can learn to deal with them effectively.

Mood Management Skills Workbook - Unit 2 - Page 16

3.

Making General Goals More Specific

- ✓ What would be different if I were approaching the goals I set in Unit 1?

- ✓ What changes would I see?

- ✓ What smaller steps are necessary in order to achieve my initial goal?

- ✓ What is giving me trouble now?

- ✓ How will I know when I'm doing better - what will happen?

Mood Management Skills Workbook - Unit 2 - Page 14

4.

Emotional Traffic Jams: the Five Roads

Triggers

Physical Responses

Thoughts

Behaviors

Feelings

Mood Management Skills Workbook - Unit 3 - Page 20

Wellness Mind - Emotional Mind

Mood Management Skills Workbook - Unit 4 - Page 35

It's the Thought That Counts!

Mood Management Skills Workbook - Unit 4 - Page 36

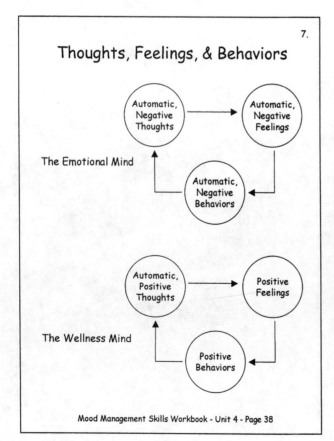

Thoughts, Feelings, & Behaviors

Mood Management Skills Workbook - Unit 4 - Page 38

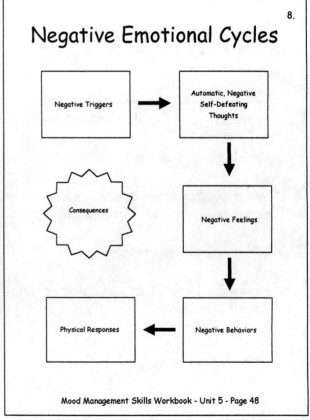

Negative Emotional Cycles

Mood Management Skills Workbook - Unit 5 - Page 48

9.

The Cycle of Depression

Triggers

Losses; Being rejected or made fun of; Being disliked; Having a chronic illness; Parents' divorce; Family problems

Thoughts

Things will never get better.
I'm a failure.
Nobody likes me.
My life is doomed.
I'm worthless.

Feelings

Hopeless; Despair; Gloom; Sad; Lonely; Rejected; Worthless; Upset; Discouraged; Somber; Disheartened

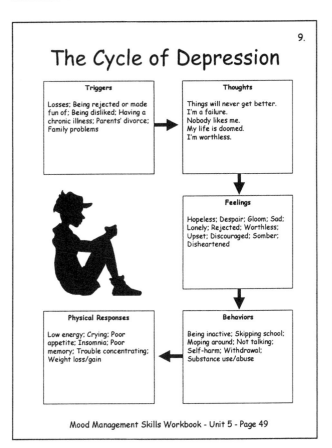

Physical Responses

Low energy; Crying; Poor appetite; Insomnia; Poor memory; Trouble concentrating; Weight loss/gain

Behaviors

Being inactive; Skipping school; Moping around; Not talking; Self-harm; Withdrawal; Substance use/abuse

Mood Management Skills Workbook - Unit 5 - Page 49

10.

The Cycle of Anger

Triggers

Being rejected or made fun of; Poverty; Emotional/physical pain; Parents' divorce; Loss; Chronic illness; Family problems

Thoughts

Everyone is out to get me.
Leave me alone.
Life is unfair.
I can't change.
I'll hurt you first.
I resent that.
I don't care.

Feelings

Irritable; Aggravated; Hopeless; Rageful; Hurt; Rejected; Hate; Annoyed; Perturbed; Exasperated; Riled up

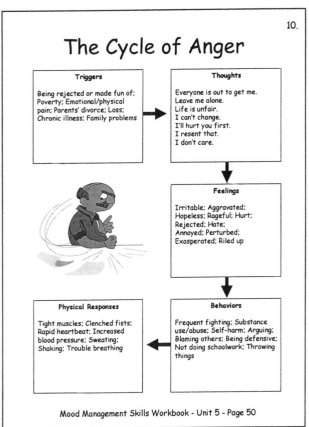

Physical Responses

Tight muscles; Clenched fists; Rapid heartbeat; Increased blood pressure; Sweating; Shaking; Trouble breathing

Behaviors

Frequent fighting; Substance use/abuse; Self-harm; Arguing; Blaming others; Being defensive; Not doing schoolwork; Throwing things

Mood Management Skills Workbook - Unit 5 - Page 50

11.

The Cycle of Anxiety

Triggers

Disasters; Life changes(moving, death); Speaking in public; Automobile accident; Chronic illness; Physical/emotional pain

Thoughts

This is really scary.
I can't handle this.
Something bad will happen.
I'm helpless.
People always make fun of me.
I'll be too embarrassed.

Feelings

Afraid; Nervous; Irritable; Confused; Panicky; Tense; Apprehensive; Helpless; Embarrassed; Shaky

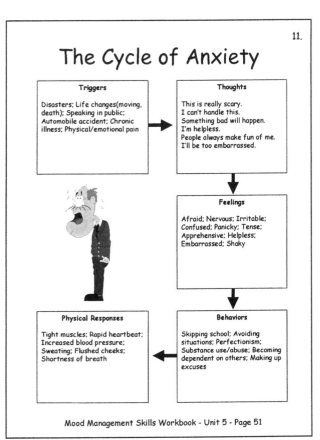

Physical Responses

Tight muscles; Rapid heartbeat; Increased blood pressure; Sweating; Flushed cheeks; Shortness of breath

Behaviors

Skipping school; Avoiding situations; Perfectionism; Substance use/abuse; Becoming dependent on others; Making up excuses

Mood Management Skills Workbook - Unit 5 - Page 51

12.

The Cycle of Low Self-Esteem

Triggers

Being picked on or made fun of; Multiple failures; Frequent, negative feedback; Many losses; Unemployment

Thoughts

I'm doomed.
I don't care.
What's the point?
I can't do anything right.
I'm helpless.
I'll fail anyway so I give up.

Feelings

Discouraged; Sad; Rejected; Lonely; Unsure; Demoralized; Overwhelmed; Frustrated

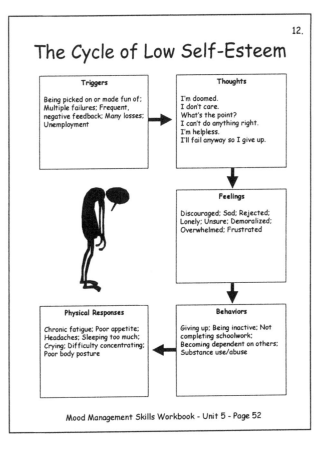

Physical Responses

Chronic fatigue; Poor appetite; Headaches; Sleeping too much; Crying; Difficulty concentrating; Poor body posture

Behaviors

Giving up; Being inactive; Not completing schoolwork; Becoming dependent on others; Substance use/abuse

Mood Management Skills Workbook - Unit 5 - Page 52

13.

Characteristics of the Emotional Mind

✓ It generates automatic, self-defeating thoughts.

✓ It wants you to believe negative things about yourself, your future, and your world.

✓ It's fast and furious. 0 to 60 in 8.2 seconds.

✓ It likes to trick you.

✓ It uses key words such as "never," "should," "always," "if/then," and "everything."

✓ It keeps you stuck in your negative emotional cycle.

✓ It often gives you the same interpretation of different triggers that, over time, causes core beliefs to develop.

Mood Management Skills Workbook - Unit 6 - Page 70

14.

Challenging: Be Your Own Mood Police

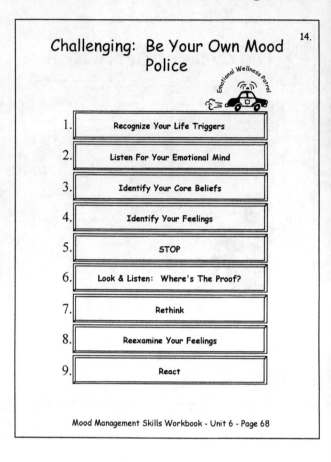

1. Recognize Your Life Triggers
2. Listen For Your Emotional Mind
3. Identify Your Core Beliefs
4. Identify Your Feelings
5. STOP
6. Look & Listen: Where's The Proof?
7. Rethink
8. Reexamine Your Feelings
9. React

Mood Management Skills Workbook - Unit 6 - Page 68

15.

Challenging: Take a Detour From Your Emotional Mind

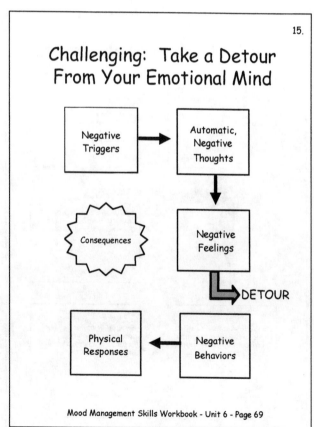

Mood Management Skills Workbook - Unit 6 - Page 69

16.

Challenging Map: The Route To The Wellness Mind

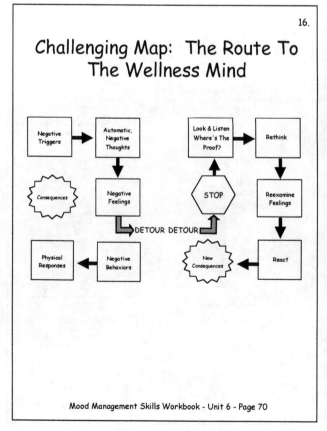

Mood Management Skills Workbook - Unit 6 - Page 70

The Thought Record

17.

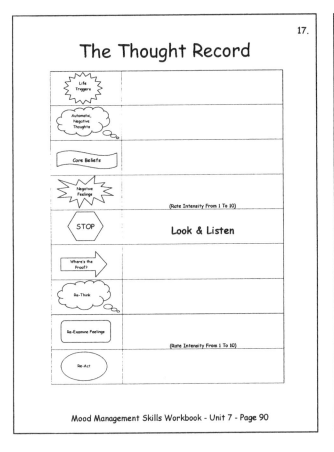

Life Triggers	
Automatic, Negative Thoughts	
Core Beliefs	
Negative Feelings	(Rate Intensity From 1 To 10)
STOP	Look & Listen
Where's the Proof?	
Re-Think	
Re-Examine Feelings	(Rate Intensity From 1 To 10)
Re-Act	

Mood Management Skills Workbook - Unit 7 - Page 90

My Action Plan...

18.

The emotion(s) that has caused me difficulty is: <u>anger</u>
This emotion has stopped me from: <u>getting along with others</u>
The small goal I would like to achieve is: <u>getting along with one person</u>

Goal(s)	Specific Steps	Evaluate Results
Get along better with my boyfriend/girlfriend	1. Listen carefully to what he/she says. 2. Buy him/her a card just because. 3. When I feel angry, count to 10 before saying anything. 4. When I feel angry, tell him/her that I need a time out. 5. When I feel angry, fill out a *Challenging Map*	1. No problems 2. He/she was surprised. 3. I had to count to 30.

Mood Management Skills Workbook - Unit 7 - Page 93

ACKNOWLEDGMENTS

The creation of the *Mood Management Skills Workbook* as well as the program for which it was designed could not have been accomplished without the support and cooperation of many people. I express my gratitude to my colleagues for their continued and enthusiastic support of the Mood Management Program. I am truly fortunate to work with these talented individuals. Their effort to move adolescents toward emotional wellness by encouraging participation in the Mood Management Program is sincerely appreciated.

I extend a special thanks to the counselors who have worked as coleaders in the Mood Management Program. Their insightful and skilled comments during group sessions as well as their willingness to share ideas for revisions of the *Skills Workbook* have played a key role in the success of the Mood Management Program. It is with the deepest respect that I thank Carolyn Arndt, Sandra Belair, John Chappell, Sheila Evjy, Maureen McGill, and Brenda Poznanski for their invaluable assistance in this endeavor.

I also thank those who have strongly supported and encouraged the use of the Mood Management Program in their schools. Sincere appreciation is extended to Nancy Duffy, Dr. P. Alan Pardy, Deborah Sadowski, and Robert Scully for their enthusiastic support of the Mood Management Program.

Lastly, I thank the adolescents who have participated in this program. Their desire to move toward emotional wellness by learning the skills promoted in this program is a process that deserves our utmost respect. I am truly honored that they have been willing to allow me to take part in their journey.

The greater part of our happiness or misery depends on our dispositions and not on our circumstances. We carry the seeds of one or the other about with us in our minds wherever we go.

Martha Washington

Introduction

Helpful Hint
As we read this section out loud, <u>underline</u> or **highlight** *things that "hit home" for you!*

Adolescence is a confusing time. Sometimes you laugh so hard that your sides ache and at other times you think things could never possibly get better. It's like a roller coaster ride. With its ups and downs, and twists and turns, sometimes things happen so fast that it makes you feel dizzy. It's a time when adults don't seem to understand a thing you're going through and your friends become more important than you ever believed imaginable.

It's a time of many firsts: your first job, your first boyfriend, your first girlfriend, and your first breakup. It is a time for major decisions: where to go to college, if to go to college, and what to do with the rest of your life. It's a time when you're caught in the middle. You're not an adult, although many of your teachers probably expect you to be. You're not a child, although your parents may still want you to be.

Caught up in all of its contradictions it is often difficult for you to understand that adolescence is simply a phase of your life. You are embroiled in its passions and caught in its web. You may act before you think and jump to conclusions before checking out the entire story. Amidst all of this confusion, you may need a "road map" that helps you find your way through this maze called adolescence.

Mood Management is a skills-building program designed to be your

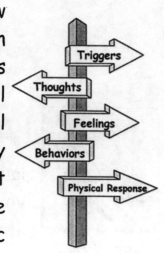

road map. It is a way for us to learn together how five important "roads" often converge to cause an emotional traffic jam. The names of these roads are shown on the street post at the right. We will talk about these roads in much more detail throughout the Mood Management Program. By using this workbook you will learn techniques that enable you to maneuver these five roads more skillfully so you can better negotiate the "traffic jams" of your adolescence. We want you to relax, have fun, and enjoy the ride.

Time Out for Discussion

1. As we read the introduction section in the workbook, what things hit home for you? What did you underline or highlight?

2. The name of one of the "five roads" is *Feelings*. What feelings do teenagers struggle with?

3. What do teenagers do to try to cope with these feelings?

4. What are some of the consequences of dealing with emotions in an ineffective way?

5. What does the word "management" mean? How does this relate to the term Mood Management?

2

A Few Ground Rules...

Mood Management is designed to be used in classroom guidance programs as well as individual or group counseling programs. When used in classroom guidance and group counseling programs, Mood Management provides adolescents with an opportunity to help one another "steer clear" of their emotional traffic jams. After all, if one person in the group is having difficulty maneuvering the five roads, it is often useful for another group member to help out by reading the road map. In this way, no group member is traveling alone. Simply stated, pack your bags and get ready to go. We are all on this journey together.

To make the journey as comfortable as possible, it's necessary to set a few guidelines for the trip. Each group member is expected to follow the guidelines very carefully. In this way, we can think of our group as "The Mood Management Team." Each member of the team will play a valuable role in the overall success of our game plan.

Group Guidelines

Let's take a few minutes to talk about each of the group guidelines listed below.

1. What does each guideline mean?

2. Are there any other guidelines you would like to add?

- Confidentiality

- Sharing

- Listening

- Dignity & Respect

- Use "I" Statements

- Be Gentle With Yourself

Unit 1: The Basic Building Blocks

Helpful Hint
As we read this unit out loud, underline or highlight things that "hit home" for you!

By now you're probably wondering, "What exactly is Mood Management?" and "How can it help me with my problems?" Let's start by answering both questions by saying that Mood Management is a skills-building program. It is designed to help you learn specific strategies that, when practiced and used effectively, pave the way to emotional wellness. It is a program that requires your commitment to learning and practicing the skills introduced in each unit of this workbook.

You can think of these skills as your tools to emotional wellness and the Mood Management Program as your toolbox. When dealing with problems such as anger, depression, anxiety, panic, stress, and low self-esteem, you can open your toolbox, select the "right" tool, and use it to manage your emotions in a better way. Certain difficulties may require many different tools as well as your commitment to "stick with it" even though it may take a while before you "get it right."

It's like riding your bike for the first time without training wheels. Do you remember what that was like? You practiced over and over until you got it right. You fell down and got up, fell down and got up, over and over again. Then one day you rode off and didn't fall. You mastered the skills required for

5

smooth sailing. Simply stated, Mood Management can help you learn the skills required for smooth emotional sailing.

Each unit in this workbook will introduce one or more skills. You will have an opportunity to learn and practice these skills in many different ways. For example, the workbook is designed to include time for general discussion about various concepts that are related to each skill. The general discussion will give you an opportunity to share your thoughts about the topic while paying special attention to how your own life relates to the issues being discussed.

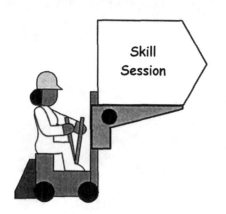

Skill Session

In addition to general discussions, each unit has a specific skill session that we complete together as a group. The introduction to each skill session is designated by the special symbol at the left. When you see it in the workbook, you will know it's time to learn and practice a new skill.

You will also practice Mood Management skills by completing weekly assignments that are similar in design to the skill sessions. By completing these assignments, you will reinforce the skills that will help you cope with emotions such as anger, depression, anxiety, and low self-esteem. Together, we will be like a construction team and the Mood Management skills will be our tools.

Our job is to clear up the traffic jams of your adolescence and build a road to emotional wellness. So, let's put on our hard hats, open the toolbox, and get started.

Emotional Wellness

1. As we read this part of Unit 1, what things hit home for you? What did you underline or highlight?

2. What questions, if any, do you have about the setup of this workbook?

3. Although we will discuss exactly what is meant by "emotional wellness" at another point in the Mood Management Program, what do you think of when you hear the term emotional wellness?

Building your road to emotional wellness requires that you think carefully about how you would like the road to look before actually starting construction. As with any construction project, building your road requires that you first develop a plan that will then serve as the blueprint for how you will actually proceed. Once the blueprint is set, you can refer to it along the way to check out how much progress you are making.

Your blueprint for emotional wellness contains the goals that you would like to achieve through your work as a member of the Mood Management group. Although your specific goals may be different from another group member's goals, the process of making the blueprint is the same. You need to think about the changes you would like to make in your life to manage your emotions more effectively. Once you have stated your goals, you can use them as a guide as you learn and practice the skills of Mood Management.

You may be wondering why we refer to "setting goals" as a skill that you learn in Mood Management. After all, setting goals seems to be a pretty easy process. Well, as it turns out, setting reasonable and obtainable goals is not as easy as it appears. In fact, there are three very important questions that you need to ask about the goals that you set before they become a part of your blueprint for emotional wellness. If you answer "yes" to each question about a particular goal, then go ahead and put it in your plan. If, however, you answer "no" to any question about a particular goal, it's time to go back to the drawing board to start over.

We will practice setting goals for emotional wellness together. On the following page, you will find a worksheet designed to help you learn the first Mood Management skill. Feel free to write in the space provided in the workbook as we complete this skill session.

Stop Here...

Complete Skill Session I: Setting Goals

The 3 Yes Rule

Here are the three questions that you need to consider to set reasonable and obtainable goals:

- Do my goals involve changing myself rather than expecting others to change?
- Do my goals involve changing things that are in my control?
- Are my goals realistic?

Examine the goals written in the following chart. Let's discuss how each goal relates to the 3 Yes Rule. When we finish discussing the examples in the chart, write one or two goals you would like to achieve.

Stated Goal	Does It Meet the 3 Yes Rule?	Can It Become Part of the Plan?
I want to learn to control my temper better.	Yes	Yes
I want to get along better with my parents.	Yes	Yes
I want everyone to like me. I'll be happy then.	No	No

 Key Points to Remember

Unit 1: The Basic Building Blocks

1. Adolescence can be a confusing time during which five important roads often converge to cause an emotional traffic jam.

2. Mood Management is a skills-building program that can help you maneuver these five roads more skillfully.

3. It is important to follow "Group Guidelines" for our group to be successful.

4. The skills that you learn in Mood Management are the tools you use to build your road to emotional wellness.

5. The first skill in Mood Management is setting reasonable and obtainable goals.

6. There are three important questions that you need to ask about each goal you set before it becomes a part of your plan.

7. Throughout the Mood Management Program, it is important for you to look back at the goals you listed in Unit 1. This way you can keep checking on your progress.

Unit 1 Assignment: Setting Personal Goals

Directions: In the space provided in the following chart, list personal goals you would like to achieve to manage your emotional difficulties more effectively. Just as you did for Skill Session I, ask yourself the three questions listed on page 9 to determine whether or not each goal meets the 3 Yes Rule.

Goal-Setting Worksheet

My Personal Goal	Does It Meet the 3 Yes Rule?	Can It Become Part of My Personal Plan?

Before We Move On

```
┌─────────────────────────────────────────┐
│              Review:                      │
│   Unit 1 - The Basic Building Blocks      │
└─────────────────────────────────────────┘
```

1. Let's take a few minutes to go over the key points on page 10 again.

2. What questions, if any, do you have about Unit 1?

3. Let's go over the weekly assignment on page 11.

```
┌─────────────────────────────────────────┐
│              Preview:                     │
│   Unit 2 - A Closer Look at Emotional Wellness │
└─────────────────────────────────────────┘
```

1. Let's take a few minutes to talk about emotional wellness in more detail. Think about someone you know who has his or her "act together."

   a. What is this person like?
   b. Describe how he or she acts when things are stressful.
   c. What emotional qualities does this person have that you admire?

2. Let's come up with our own definition of emotional wellness.

3. Let's move on to Unit 2 and take "A Closer Look at Emotional Wellness."

Unit 2: A Closer Look at Emotional Wellness

Helpful Hint
*As we read this section out loud, <u>underline</u> or **highlight** things that "hit home" for you!*

Before we begin the Skill Session for Unit 2, let's take a closer look at what we mean by the term "emotional wellness" and consider why this concept is so important to the work that you do as a member of the Mood Management group. You discovered from the previous unit that the goal of the Mood Management Program is to help you learn specific skills that you can use to build your road to emotional wellness. You even started to construct that road by setting goals you want to achieve through your participation in the Mood Management Program.

Although you have started your journey toward emotional wellness, you may still be asking yourself, "What exactly is meant by this?" and "How will developing emotional wellness help with the difficulties I encounter during my adolescence?" One of the most useful ways to get a better picture in your mind about what we mean by emotional wellness is to ask the question, "What does an emotionally well adolescent look like?" To answer this question, let's take a look at the following scenario about Edward Wellbeing, a 10th-grade student at Fountain Spring High School who completed the Mood Management Program and developed the skills necessary to complete the construction of his road to emotional wellness.

Prior to his participation in the Mood Management Program, Edward could best be described as an angry, unhappy person who was not doing very well in school. His grades were dropping, he was not getting along with teachers, and his friendships seemed to be dwindling. Although this made Edward feel very upset "deep down inside," he would often tell people that he didn't care about these things. In fact, "I don't care" started to be Edward's response to the difficulties he encountered from day to day.

For a time, this response seemed to work for Edward. He would tell people that he didn't care, and they would leave him alone. Edward believed that if people would just leave him alone, he wouldn't have to think about his problems. In fact, he tried his very best to ignore the difficulties he was experiencing with the hope that they would just go away. After awhile, however, Edward discovered that this way of coping with his problems was not working. He was becoming more and more unhappy, and his anger seemed to be mounting. He couldn't seem to find any way to feel better and started to believe that he would never be happy.

Frustrated and feeling hopeless, Edward decided to join the Mood Management Program. He wasn't convinced that it would really help him feel better, but a friend of his had completed the program and encouraged Edward to give it a try. During his participation in the Mood Management group, Edward learned a great deal about himself and the choices he had been making.

Edward learned that he hadn't been making very good choices in response to his difficulties. Although he had believed that making these choices was the only way to cope with his problems, Edward discovered that he had relied on them because he didn't know of a better way to go about problem solving. Through participation in the

Mood Management Program, Edward learned that he had many more options when faced with a problem situation. He also learned that he wasn't the only one struggling with the emotional challenges of adolescence and found comfort in talking with other teenagers about this.

As Edward continued on his journey to emotional wellness, he began to change. He realized that his "I don't care" philosophy wasn't very useful, so he discarded it. He became willing to look closely at his problems and to acknowledge that he did care about how he acted when things got tough for him. Furthermore, he made a commitment to develop the skills that would enable him to achieve emotional wellness. Through hard work and effort, Edward was able to "get his act together," and he managed his problems much more effectively as a result.

Time Out for Discussion

1. As we read this part of Unit 2, what things hit home for you? What did you underline or highlight?

2. What were the feelings that Edward was struggling with?

3. What did Edward do to try to cope with these feelings?

4. What do you think about the way Edward tried to cope with his feelings? Have you done some of the same things to cope with your feelings?

5. Edward didn't think that the Mood Management Program could really help him, but he decided to try it anyway. Do any of you have similar thoughts? Let's discuss this.

6. The following acronym sums up our concept of emotional wellness. When we ask if you have been "minding your image," we are referring to this acronym, not to the way you look or dress. Let's take a few minutes to talk about this.

I do care.

Managing problems effectively is a plus.

Awareness is important.

Go for it - it's worth the effort!

Emotions - we all have them and can learn to deal with them effectively.

Now that you have a better picture of what we mean by emotional wellness, let's move on. Remember, the Mood Management Program is your toolbox. Let's open it and learn how to use the next tool as you continue your journey toward emotional wellness.

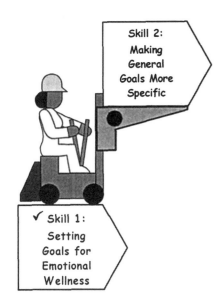

At this point, it's important to reexamine the goals you set in Unit 1. Many times when we first set a goal, it needs to be redefined because it is too general. Redefining initial goals that are too general requires that you break them down into smaller, more specific goals that can be observed and measured as you continue to build your road to emotional wellness. To make your initial goals more specific, ask the following questions:

- What would be different if I were approaching the goals I set in Unit 1? What changes would I see?

- What smaller steps are necessary to achieve my initial goal?

- What is giving me trouble now? How will I know when I'm doing better - what will happen?

Let's use these questions as a guide to completing the chart on the following page. Look at the example that is given in the chart. How is the initial goal different from the revised, more specific goals?

Choose one of your initial goals from Unit 1, and write it in the first column on the chart. Then fill in the second and third columns.

Stop Here...

Complete Skill Session II:
Making General Goals More Specific

General Goal From Unit 1	What Changes Would I See?	Revised, More Specific Goal
I want to feel better about myself.	I would recognize my good points. I would stop "beating myself up" when I make a mistake.	To recognize and state my strengths. To accept that it is "OK" to make a mistake.

 Key Points to Remember...

Unit 2: A Closer Look at Emotional Wellness

1. The goal of Mood Management is to help you develop skills that will enable you to deal with your problems more effectively.

2. Mind your __IMAGE__ !! Review the acronym on page 16 to remind yourself about emotional wellness.

3. You can think of emotional wellness as getting "your act together." It takes specific skills to do this. These skills can be learned.

4. The second skill in Mood Management is making your initial goals more specific by breaking them down into smaller goals that are observable and measurable.

Unit 2 Assignment: Setting More Specific Goals

Directions: In the space provided in the following chart, list your initial goals from Unit 1. Then list what changes you would notice if you were making progress toward each goal. In the last column, write a revised, more specific goal.

General Goal From Unit 1	What Changes Would I See?	Revised, More Specific Goal

Before We Move On

> **Review:**
> Unit 2 - A Closer Look at Emotional Wellness

1. Let's take a few minutes to review the key points on page 19.

2. What questions, if any, do you have about Unit 2?

3. Who "minded their IMAGE" this week?

4. Let's go over the weekly assignment on page 20.

> **Preview:**
> Unit 3 - The Emotional "Traffic Jams" of Adolescence

1. Let's take a few minutes to talk about traffic jams in more detail.

   a. How many of you have been stuck in a traffic jam?
   b. What is it like to be in a traffic jam?
   c. What feelings do you experience?
   d. What are some of the things that people do when they are caught in a traffic jam?

2. How does being stuck in a real traffic jam relate to the title of Unit 3?

3. How many of you have felt like you were stuck emotionally with no way out? What is this like?

Unit 3: The Emotional "Traffic Jams" of Adolescence

Helpful Hint
As we read this unit out loud, <u>underline</u> or **highlight** *things that "hit home" for you!*

Congratulations! By learning the first two skills found in your Mood Management toolbox, you have already begun the construction of your road to emotional wellness. Setting initial goals and then redefining them has provided you with a picture of how you want your specific road to look. As you continue on your journey, you can review your redefined goals and determine how well you are moving toward them as you learn the skills introduced in each additional Mood Management unit.

In Unit 1, we introduced the notion that adolescents may have difficulty achieving emotional wellness because of certain "traffic jams" that block their journey. These traffic jams are caused by five roads that often converge to cause emotional havoc for adolescents. To move forward on their journey to wellness, adolescents must be able to clear up the traffic jams that are causing them emotional difficulty. Before they set out to clear up these traffic jams, however, adolescents have to be able to read a road map that shows exactly where, how, and why these five roads converge.

The next tool in your Mood Management toolbox is that road map. By studying it carefully and learning all about the five roads, you will be able to start clearing up your own emotional traffic jam. Keep in

mind that good traffic officers know the roads! They know how roads meet and intersect as well as which road would be best to clear first when a traffic jam does occur. For your journey to continue smoothly, you will act as your own traffic officer. So, let's study the road map, learn all about the five roads, and get that traffic moving!

As you know, Mood Management is designed to assist you in learning skills that will enable you to better negotiate the emotional ups and downs of your adolescence. One of the most important things to learn in Mood Management is that the strong feelings associated with these emotional ups and downs do not happen in isolation. In fact, strong feelings are intertwined with four very important factors that must be closely examined as you continue on your path to emotional wellness. Strong feelings and the other four factors make up the five roads that cause emotional traffic jams.

Although it often seems as if there is nothing else happening when you experience a strong emotion, it is important to learn that there are quite a few things going on every time you feel stuck in your anger, sadness, depression, anxiety, or low self-esteem. Strong feelings can be overwhelming, and they often blur our vision. However, if you look closely enough you will be able to see the relationship that exists among

➡ triggers
➡ thoughts
➡ feelings
➡ behaviors
➡ physical responses

> These are the five roads. Study them carefully to learn the skills required to better manage your emotions.

To help you better imagine what we mean by the five roads of emotional traffic jams, look at the following diagram. Let's take a few minutes to talk about what this diagram means.

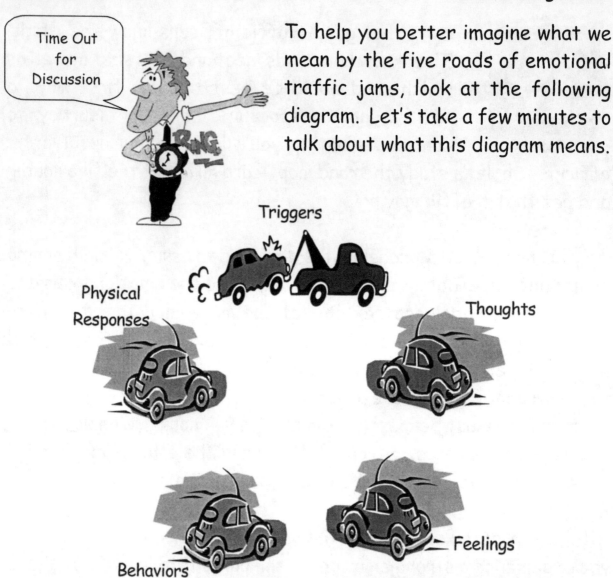

1. What do you notice when you look at the diagram?

2. How do you feel when you're stuck in a real traffic jam? What are some of your thoughts? What do you do? How does your body respond? How does this relate to an emotional traffic jam?

3. In the center of the diagram, write a number from 1 to 10 that best describes how stuck you feel emotionally (10 is the most). Under the number, write the name of the emotion(s) that has you stuck.

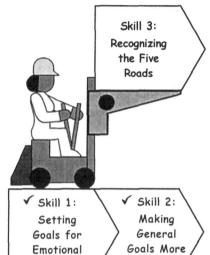

To help show you how these five roads intersect to cause emotional havoc for many adolescents, let's revisit Edward Wellbeing and take a closer look at what his life was like before he joined the Mood Management Program.

As you remember, Edward was stuck in an emotional traffic jam filled with anger and sadness. Prior to attending Fountain Spring High School, Edward had gone to school in New York City. In fact, he had spent his entire life in New York. He had many friends that he had known from the time he started school in the first grade. These were his best friends, and Edward thought that he would graduate from high school with them. Then came his parents' divorce and the move to Fountain Spring.

Helpful Hint
As we read this section out loud, underline or **highlight** things that "hit home" for you!

Edward didn't like Fountain Spring. He told himself that it couldn't compare to New York City and that he would never be able to make friends like the ones he had back home. He thought, "I don't fit in here, anyway, so why bother even trying to make friends?" Edward was angry with his parents. "Why did they have to get a divorce?" "Why did he have to move away from his home and friends?"

As time went on, Edward became more isolated at Fountain Spring High School. He was bored in school and didn't like many of his teachers. He skipped school a few times because he disliked it so much. Edward got caught, however, and the principal called Mrs. Wellbeing. She was very disappointed with Edward and told him so. Edward felt terrible. He hadn't meant to upset his mother. He thought, "No matter what I do it isn't right. I should just give up trying. Nothing really matters."

As time went on, Edward became more and more angry. He was unhappy and found little enjoyment in anything. He wasn't sleeping well, and most of the time he wasn't in the mood to eat anything. In fact, Edward began to lose weight. It was as if Edward had fallen into a huge hole in the ground. He couldn't seem to find his way out. He was caught in a cycle of anger and depression that continued to spiral out of control until he learned the skills required to manage his strong emotions effectively.

Now that you know more about Edward's traffic jam, let's see if you can find the five roads in his story. Use the chart on the next page to fill in the information about each of Edward's five roads. Remember, the names of the five roads are

Stop Here...

Complete Skill Session III:
Recognizing the Five Roads

Triggers: _parents' divorce, moving to Fountain Spring, leaving his good friends behind, not being able to graduate from high school in New York..._

Thoughts: _I don't fit in here. Why bother to try and make friends? Why did my parents have to get a divorce? No matter what I do it isn't right. I should just give up trying. Nothing really matters._

Feelings: _angry, bored, unhappy, depressed, lonely, hopeless, frustrated..._

Behaviors: _isolating himself from others, skipping school, giving up, not eating..._

Physical Responses: _weight loss, not sleeping well..._

Key Points to Remember...

Unit 3: The Emotional Traffic Jams of Adolescence

1. There are five roads that converge to cause emotional havoc for adolescents.

2. Although strong feelings may appear to happen in isolation, they are closely related to thoughts, behaviors, triggers, and physical responses.

3. By studying the five roads, you will learn skills to help manage your strong emotions more effectively.

4. Each of the five roads intersects with the others.

5. Small changes in one road cause changes in the other four.

Unit 3 Assignment:
Recognizing The Five Roads

Directions: Being able to recognize the difference between feelings, thoughts, behaviors, triggers, and physical responses is critical to the work you do in the Mood Management Program.   Identify each of the following items as a feeling, thought, behavior, trigger, or physical response.

1.   No matter what I do it isn't right   _____*thought*_____

2.   Irritated   _____*feeling*_____

3.   Failing a test   _____*trigger*_____

4.   No one likes me   _____*thought*_____

5.   Angry   _____*feeling*_____

6.   Blaming others   _____*behavior*_____

7.   Embarrassed   _____*feeling*_____

8.   I bet other kids did better on the test   _____*thought*_____

9.   Rapid heart beat   _____*physical response*_____

10.   Clenched fist   _____*physical response*_____

11.   Who cares?   _____*thought*_____

12.   Parents' divorce   _____*trigger*_____

13.   Getting stuck in traffic   _____*trigger*_____

14. Nervous _____ *feeling*

15. Moving to a new city _____ *trigger*

16. Sweating _____ *physical response*

17. Being picked on or made fun of _____ *trigger*

18. I'm always the one who gets yelled at _____ *thought*

19. Skipping school _____ *behavior*

20. Sad _____ *feeling*

21. Using drugs or alcohol _____ *behavior*

22. Throwing things across the room _____ *behavior*

23. Withdrawing from friends _____ *behavior*

24. Difficulty breathing _____ *physical response*

25. I can't take this anymore _____ *thought*

26. Nothing will ever change _____ *thought*

27. Lonely _____ *feeling*

28. Crying _____ *physical response*

29. Headaches _____ *physical response*

30. I can't really change _____ *thought*

Before We Move On

> **Review:**
> Unit 3: The Emotional "Traffic Jams" of Adolescence

1. Let's take a few minutes to review the key points on page 28.

2. What questions, if any, do you have about Unit 3?

3. Let's go over the assignment on pages 29 and 30.

> **Preview:**
> Unit 4: Clearing Up Those Traffic Jams

1. Let's review the diagram on page 24.

2. Which road are you stuck on?

3. Which road do you think we should clear first to get traffic moving?

Unit 4: Clearing Up Those Traffic Jams

Helpful Hint
As we read this section out loud, <u>underline</u> or **highlight** *things that "hit home" for you!*

As you continue to study your map of the five roads that cause emotional traffic jams, you will discover that getting traffic rolling as quickly and smoothly as possible depends on which of the five roads you choose to clear first. Although each road intersects with and causes changes in the other four, there is one road that acts as a superhighway. When it is blocked, your emotional traffic flow can be caught up for a very long time. When this happens, you are not able to mind your IMAGE, and it is likely that you feel overwhelmed by your emotions. This, in turn, makes it very difficult to move toward the goals that you want to achieve for emotional wellness. To move forward, then, you need to clear your traffic jam. To do this as quickly and smoothly as possible, you will find that it is best to start with the "thought superhighway."

Have you ever heard the expression, "It's the thought that counts"? How about, "Thinking makes it so"? Here's still another, "I think, therefore I am." Each of these expressions suggests that thoughts play a powerful role in how we experience a particular situation. This is especially true for the emotional ups and downs of adolescence. The thoughts that you experience when you get caught up in an emotional traffic jam play a powerful role in determining whether or not you will remain stuck in that traffic jam. There are two types of thoughts which travel along your thought superhighway.

One type of thought stops all traffic, whereas the other gets things moving. Learning about these thoughts and becoming familiar with the influence each type has over your ability to clear up your traffic jam is a critical skill that must be learned for you to be able to continue your journey toward emotional wellness.

You know that you're in an emotional traffic jam when you feel stuck in emotions such as anger, frustration, loneliness, depression, anxiety, or low self-esteem. Emotional traffic jams make you feel overwhelmed, and you may think that things will never get better. During these traffic jams, your thought superhighway is hard at work generating ideas that we call "traffic blockers." These types of ideas or thoughts come from your "emotional mind," and as long as they remain unchecked your traffic remains stuck.

When you are effectively managing your emotions, however, your traffic is flowing. Your thought superhighway is generating ideas that we call "traffic movers," which come from your "wellness mind." Although things may be difficult, you don't feel stuck or overwhelmed by your emotions. Learning to get your wellness mind up and running is a skill that you learn in Mood Management. Although all adolescents have a wellness mind that is capable of generating traffic-moving thoughts, it is not unusual if you need a little help in getting your wellness mind going. Once it's on the move, however, you will have the skills to feel better, change behavior, and move toward emotional wellness.

To clear up your emotional traffic jam, you need to be able to identify the types of thoughts you have and understand the impact that these thoughts have on your feelings and behaviors. Thoughts generated by the wellness mind cause different feelings than those generated by the thoughts from the emotional mind. Thoughts from the emotional mind tend to be automatic, negative thoughts that keep you stuck in your emotional traffic jam. Thoughts generated from the wellness mind, however, are positive and realistic thoughts that take time to learn. Each type of thought "drives" different types of feelings. Negative thoughts from the emotional mind drive negative feelings. Positive, realistic thoughts from the wellness mind drive positive feelings.

The emotional mind tends to be very powerful. It reacts very quickly in stressful situations, producing thoughts that are negative and self-defeating. You know that you are in your emotional mind when you feel overwhelmed by emotions and can't seem to feel better no matter how hard you try. When you feel overwhelmed by depression, anger, anxiety, or low self-esteem, be aware that your emotional mind is very active. In fact, it's going full force doing its job, which unfortunately makes you feel miserable.

1. As we read this part of Unit 4, what things hit home for you? What did you underline or highlight?

Time Out
for
Discussion

2. Let's take a few minutes to look at the diagram on page 35. It will give us a better idea about what we mean by the emotional mind and wellness mind and the types of thoughts generated by each.

a. What do you notice about this diagram?

b. What types of thoughts are found in the emotional mind?

c. What types of thoughts are found in the wellness mind?

d. How many of you have been in your emotional mind today?

e. What types of thoughts have you had?

f. How many of you have been in your wellness mind today?

g. What types of thoughts have you had?

3. Here's another example to help show how the thoughts generated by the emotional mind and the wellness mind influence our feelings and behaviors. Suppose that two students each get results from a recent test and each student has failed the test. In the exact same situation (trigger), one student's emotional mind is operating while the other student's wellness mind is operating.

   a. Which student is in his emotional mind? How can you tell?
   b. How is each student feeling?
   c. How is each student likely to behave?
   d. What influence do our thoughts have over our feelings and behaviors?
   e. Is it the trigger or the emotional mind's interpretation of the trigger that gets us stuck emotionally?

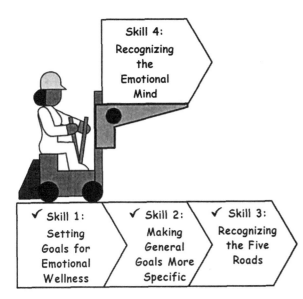

The thoughts generated by the wellness mind and emotional mind have a powerful influence over the types of feelings that adolescents experience. These thoughts and feelings, in turn, cause adolescents to behave in various ways depending on whether the initial thought was from the wellness mind or the emotional mind.

Understanding the connection between thoughts, feelings, and behaviors is very important in your continued journey to emotional wellness. Although each of the five roads intersects the others, it is the thought superhighway that really paves the way for your feelings and behaviors in any given situation. Remember, thoughts generated by your wellness mind are positive and learned. They, in turn, cause or "drive" positive feelings and behaviors. Thoughts generated by your emotional mind, however, are automatic and negative. They, in turn, cause or drive negative feelings and behaviors.

The diagram on page 38 will help you visualize the relationship between thoughts, feelings, and behaviors. Study it carefully so you become familiar with the notion that your emotional mind generates thoughts that keep you stuck in your emotional traffic jam, whereas your wellness mind generates thoughts that get your traffic moving.

Let's take a few minutes to discuss the diagram, and then we'll move on to the skill session on page 39.

Thoughts, Feelings, & Behaviors

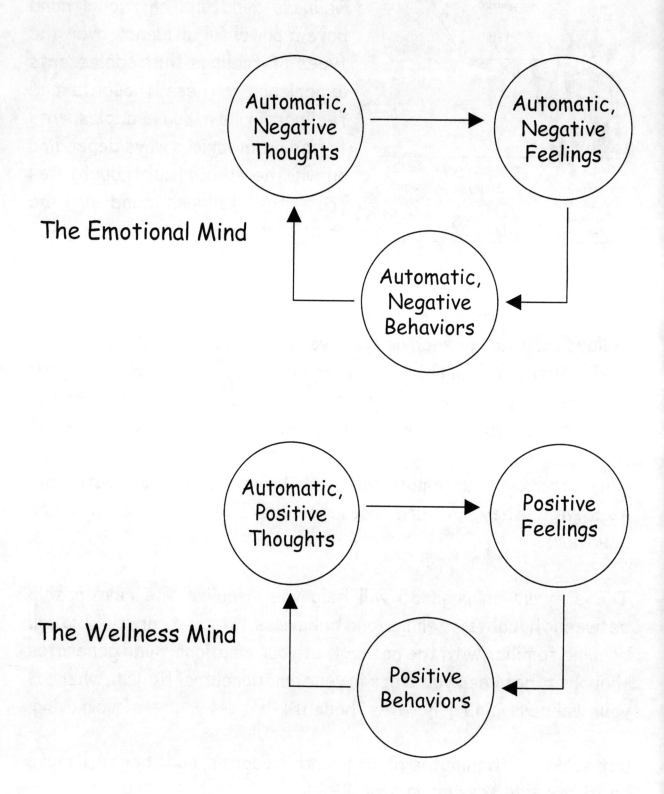

The Emotional Mind

The Wellness Mind

Stop Here...

Complete Skill Session IV:
Thoughts, Feelings, & Behaviors

Thought	Emotional or Wellness Mind?	Possible Feeling?	Possible Behavior?
This class is so hard. I know I'm going to fail it.	emotional mind	discouraged, hopeless, frustrated...	giving up, dropping the class...
I probably won't make the team. There are so many kids that are better than I am.	emotional mind	frustrated, hopeless, discouraged, angry...	not trying out for the team, giving up...
I need to try harder next time. I want to do the best that I can.	wellness mind	determined, hopeful, unshakable, assured, encouraged...	practice more, staying involved...
What's the worst thing that can happen? If I ask her to dance, she might say no. But, she might say yes. I won't know unless I ask her.	wellness mind	hopeful, nervous, excited, optimistic, encouraged...	asking her to dance...
Everyone here is so unfriendly. I wish we had never moved to this town.	emotional mind	sad, angry, lonely, discouraged, regretful...	withdrawing from others, skipping school, not completing assignments, substance use...
I'm a little anxious about being here, but there are so many new kids to meet.	wellness mind	excited, nervous, encouraged...	trying to meet others, participating in activities...

   Key Points to Remember...

Unit 4: Clearing Up Those Traffic Jams

1. Thoughts, feelings, and behaviors are closely related.

2. When you feel stuck in an emotional traffic jam, your emotional mind is busy at work generating automatic, negative thoughts that, in turn, drive your negative feelings and behaviors.

3. Learning to recognize how thoughts drive feelings and behaviors can lead to moving forward on your road to emotional wellness.

4. Mood Management is more than just "positive thinking." It requires that you learn specific skills that will enable you to better manage your thoughts, feelings, and behaviors.

Unit 4 Assignment: Thoughts, Feelings, & Behaviors

Directions: Complete the following chart just as you did for Skill Session IV. Next to each thought, write whether it comes from the emotional or wellness mind. Write a word for the possible feeling it would generate. Next to each feeling, write a word for the possible behavior that would be generated.

Thought	Emotional or Wellness Mind?	Possible Feeling?	Possible Behavior?
No one respects me.	emotional mind	sad, angry, lonely, discouraged...	withdrawing from others, substance use, skipping school...
I will never be successful.	emotional mind	discouraged, mad, overwhelmed...	giving up, not completing schoolwork, withdrawing...
I do care.	wellness mind	hopeful, optimistic, encouraged...	staying involved, eating right, doing schoolwork...
Things are difficult, but I will keep trying.	wellness mind	hopeful, assured, determined, encouraged...	trying harder, hanging in there, completing assignments...
I can't please everyone all the time, and that's O.K.	wellness mind	confident, assured, relaxed...	trying to do the best, accepting limitations, staying involved...
Other people are always disappointed in me.	emotional mind	sad, nervous, overwhelmed, unhappy, sad...	giving up, withdrawing from others, skipping school, substance use...
I am an important person in this world.	wellness mind	confident, happy, relaxed...	being involved, completing schoolwork...
My life will never get any better.	emotional mind	sad, discouraged, lonely, dejected...	giving up, withdrawing from others, substance use...
I'm a good friend to people.	wellness mind	happy, confident, proud...	being involved, listening to others, caring for others...

Before We Move On

Review:
Unit 4: Clearing Up Those Traffic Jams

---

1. Let's take a few minutes to review the key points on page 40.

2. What questions, if any, do you have about Unit 4?

3. Let's go over the weekly assignment on page 41.

Preview:
Unit 5: The Cyclones of the Emotional Mind

---

1. Let's take a few minutes to talk about cyclones in more detail.

   a. What is a cyclone?
   b. What does it do to everything in its path?
   c. How does the concept of a real cyclone relate to the title of Unit 5?
   d. What do the thoughts from your emotional mind do to everything in their path?

2. There is no way to stop a real cyclone. It has to run its course before it loses momentum and dies out. Fortunately, your emotional mind doesn't have to continue running its course. Let's move on and figure out how to get it to stop spinning.

Unit 5: The Cyclones of the Emotional Mind

Helpful Hint
As we read this unit out loud, underline or highlight things that "hit home" for you!

Before we introduce the information in this unit, let's take a few minutes to review what you have learned so far. After all, you have dedicated a significant amount of time and energy to learning the skills introduced in Units 1 through 4, and at this point it's important for you to slow down, catch your breath, and take an inventory of all you have accomplished. So, let's step back and take a look.

Although you have learned many important concepts from the previous units, perhaps the most important deals with the notion of emotional wellness. Let's face it. This is why you're here! You have chosen to learn how to manage your difficult emotions more effectively and, no matter how you look at it, your decision to do this is nothing to take lightly. As a matter of fact, it's admirable. Unlike the many people who live their lives stuck in an emotional traffic jam, you have decided to get a better grip on your emotional ups and downs to live a happier and more fulfilling life.

As you know, emotionally well adolescents do this by minding their IMAGE. They care about themselves, manage their emotions effectively, and have awareness that dealing with emotions in an ineffective manner keeps them stuck in an emotional traffic jam.

They also know, however, that emotional wellness can be learned, so they go for it! They practice the skills required to clear up their traffic jams so they can move toward the goals they want to achieve. Thus, emotionally well adolescents get their traffic moving.

To do this, emotionally well adolescents frequently examine their map of the five roads. They know that emotional havoc occurs when the five roads converge, causing them to feel overwhelmed, confused, and stuck. Since emotionally well adolescents choose to get their traffic moving, they use their road map as a guide to getting "unstuck." In fact, they act as traffic officers. As such, they know the names of the five roads: Thoughts, Feelings, Behaviors, Triggers, and Physical Responses. In addition, they realize that the best way to get their traffic moving is to start with the thought superhighway.

Your thought superhighway generates two types of thoughts, traffic blockers and traffic movers. Traffic blockers, which come from your emotional mind, are automatic, negative thoughts that drive negative feelings and negative behaviors. Traffic movers, however, are realistic, positive thoughts that drive positive feelings and positive behaviors. They come from your wellness mind. Learning to shift from your emotional mind to your wellness mind is an important tool found in your Mood Management toolbox. You will learn how to use this tool in future units.

Time Out for Discussion

1. Let's take a few minutes to talk about all that you've learned so far. What key words did you underline or highlight?

2. The following terms represent the concepts that you have learned so far and, quite frankly, this is a lot of information to have mastered in only a short period of time. You've done it, however. Congratulations. Your hard work is indeed something to be proud of.

Emotional wellness
The five roads
Feelings
Physical responses
Traffic blockers
Awareness
Thoughts
Triggers
Wellness mind
Positive, realistic thoughts
Emotionally well adolescents

Emotional mind
Automatic, negative thoughts
Emotional traffic jams
Thought superhighway
Setting goals
IMAGE
Behaviors
Traffic movers
Traffic blockers
Managing emotions effectively

3. What are your thoughts about all you have learned so far?

4. Which concepts have been the most meaningful to you?

Way To Go!

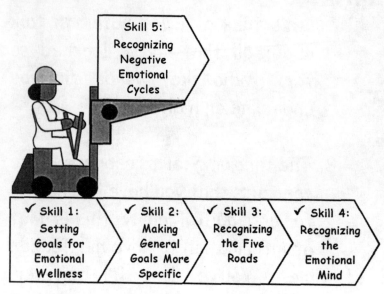

Now, let's move on and examine the different kinds of traffic jams caused by the automatic, negative thoughts of the emotional mind. As you know, these thoughts are called traffic blockers. They come on like a cyclone, fast and furious, without any warning. They can destroy everything in their path and leave you feeling miserable.

In Unit 4, you learned that emotional traffic jams are caused by the automatic, negative thoughts generated by your emotional mind. These negative thoughts, brought on and reinforced by certain triggers, are referred to as self-defeating thoughts because they make us feel very badly about ourselves. In fact, you can say that these negative, self-defeating thoughts are your own worst enemy. They cause you to doubt yourself, make you believe that things will never get better, and ultimately trap you in the overwhelming feelings of depression, anger, anxiety, and low self-esteem. These are painful feelings that are experienced in direct response to the disturbing, negative, automatic thoughts of your emotional mind.

To feel better, then, you first need to recognize the cycles of depression, anger, anxiety, and low self-esteem. Once you recognize these common cycles, you then need to identify your own five roads to determine what emotional cycle has you trapped. When you have done this, you will have the skills necessary to learn how to become free of your self-defeating thoughts that will, in turn, free you from your emotional traffic jam.

Time Out
for
Discussion

Later, we will study these cycles very carefully so you can determine which one(s) has you trapped. Before we do that, however, take a look at the diagram on page 48. It is a general diagram of the stages found in all negative emotional cycles.

1. What do you notice about this diagram? Where have you seen something like this before?

2. Notice that the cycle starts with a trigger. Remember, triggers are things that happen that we don't have control over. Negative triggers are things that we wouldn't choose to have happen. A trigger can be something that happened a long time ago (lifelong trigger) or something that happened recently.

   a. Negative triggers are like fuel for the emotional mind. What happens when the emotional mind starts to burn fuel?

   b. The emotional mind has a storage area for triggers that we call a "trigger bank." The trigger bank not only stores memories of the negative event but also stores the emotional mind's interpretation of the event. What do you think this does to the automatic, negative thoughts from the emotional mind?

   c. Name some common triggers that teenagers experience.

d. Let's take a few minutes to discuss this general diagram of all negative emotional cycles in more detail and then move on to the cycles of depression, anger, anxiety, and low self-esteem.

Negative Emotional Cycles

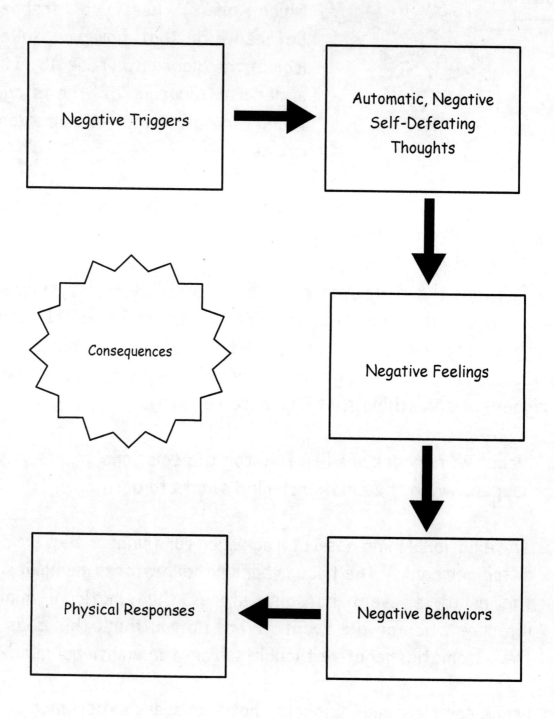

The Cycle of Depression

Triggers

Losses; Being rejected or made fun of; Being disliked; Having a chronic illness; Parents' divorce; Family problems

Thoughts

Things will never get better.
I'm a failure.
Nobody likes me.
My life is doomed.
I'm worthless.

Feelings

Hopeless; Despair; Gloom; Sad; Lonely; Rejected; Worthless; Upset; Discouraged; Somber; Disheartened

Physical Responses

Low energy; Crying; Poor appetite; Insomnia; Poor memory; Trouble concentrating; Weight loss/gain

Behaviors

Being inactive; Skipping school; Moping around; Not talking; Self-harm; Withdrawal; Substance use/abuse

The Cycle of Anger

Triggers

Being rejected or made fun of;
Poverty; Emotional/physical
pain; Parents' divorce; Loss;
Chronic illness; Family problems

Thoughts

Everyone is out to get me.
Leave me alone.
Life is unfair.
I can't change.
I'll hurt you first.
I resent that.
I don't care.

Feelings

Irritable; Aggravated;
Hopeless; Rageful; Hurt;
Rejected; Hate;
Annoyed; Perturbed;
Exasperated; Riled up

Physical Responses

Tight muscles; Clenched fists;
Rapid heartbeat; Increased
blood pressure; Sweating;
Shaking; Trouble breathing

Behaviors

Frequent fighting; Substance
use/abuse; Self-harm; Arguing;
Blaming others; Being defensive;
Not doing schoolwork; Throwing
things

The Cycle of Anxiety

Triggers

Disasters; Life changes (moving, death); Speaking in public; Automobile accident; Chronic illness; Physical/emotional pain

Thoughts

This is really scary.
I can't handle this.
Something bad will happen.
I'm helpless.
People always make fun of me.
I'll be too embarrassed.

Feelings

Afraid; Nervous; Irritable; Confused; Panicky; Tense; Apprehensive; Helpless; Embarrassed; Shaky

Physical Responses

Tight muscles; Rapid heartbeat; Increased blood pressure; Sweating; Flushed cheeks; Shortness of breath

Behaviors

Skipping school; Avoiding situations; Perfectionism; Substance use/abuse; Becoming dependent on others; Making up excuses

The Cycle of Low Self-Esteem

Triggers

Being picked on or made fun of;
Multiple failures; Frequent,
negative feedback; Many losses;
Unemployment

Thoughts

I'm doomed.
I don't care.
What's the point?
I can't do anything right.
I'm helpless.
I'll fail anyway so I give up.

Feelings

Discouraged; Sad; Rejected;
Lonely; Unsure; Demoralized;
Overwhelmed; Frustrated

Physical Responses

Chronic fatigue; Poor appetite;
Headaches; Sleeping too much;
Crying; Difficulty concentrating;
Poor body posture

Behaviors

Giving up; Being inactive; Not
completing schoolwork;
Becoming dependent on others;
Substance use/abuse

Stop Here...

Complete Skill Session V - Parts A & B

Directions: This skill session has two parts. Part A, on page 54, is called Negative Emotional Cycles. Next to each term in the chart on page 54, identify to which of the five roads it is likely to belong, identify the negative emotional cycle to which it could apply, and then check whether or not you have experienced the term being described.

Part B, on page 55, is called Mapping Out Your Own Emotional Cycle. This is an opportunity for you to map out the five roads of the cycle that has you stuck emotionally. Fill in each box of the map, starting with a trigger. The trigger can be something current or it can be a lifelong trigger. When describing the trigger, be general. It is not necessary to go into the specific details of the event. For example, you could use the general term "multiple losses" rather than describing specific details of the situation.

In the remaining boxes, be as specific as possible. Get in touch with the thoughts being generated by your emotional mind. Remember, the emotional mind's interpretation of the trigger includes automatic, negative thoughts about you, how you view life, and how you envision the future.

After you have completed filling in the boxes of your road map, assign a number from 1 to 10, with 10 being the highest, that best depicts the intensity of the feelings associated with this emotional traffic jam. Also, list the consequences (what happened) as a result of this emotional cycle.

Stop Here...

Complete Skill Session V - Part A
Negative Emotional Cycles

Term	Which of the Five Roads?	Possible Cycle? *more than one may apply*	Place a Check in This Column if You Have Experienced This
Sadness	Feeling	Depression	✔
Parents' divorce	*trigger*	*depression*	
I'm no good at this.	*thought*	*low self-esteem*	
Things will never get any better.	*thought*	*depression*	
Rejected	*feeling*	*low self-esteem*	
I can't do anything right.	*thought*	*depression*	
Self-harm	*behavior*	*depression*	
This is totally unfair.	*thought*	*anger*	
Irritated	*feeling*	*anger*	
Swearing at others	*behavior*	*anger*	
You're the one to blame for this mess!	*thought*	*anger*	
I have to get out of here.	*thought*	*anxiety*	

Skill Session V - Part B: Mapping Out Your Own Emotional Cycle

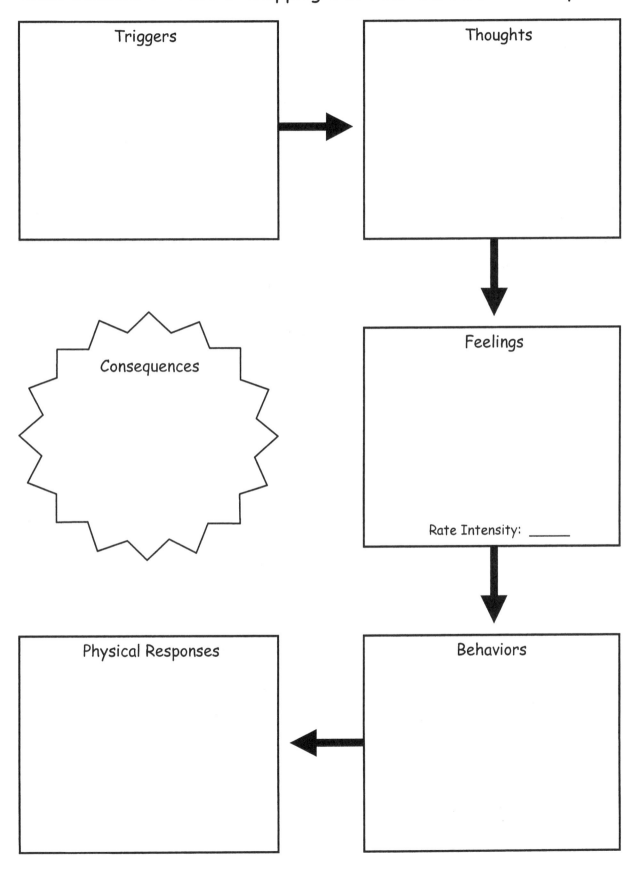

Triggers

Thoughts

Consequences

Feelings

Rate Intensity: _____

Physical Responses

Behaviors

Key Points to Remember...

Unit 5: The Cyclones of the Emotional Mind

1. You know you are stuck in an emotional traffic jam when you experience overwhelming feelings of depression, anger, anxiety, low self-esteem, or all these.

2. Emotional traffic jams are caused when your emotional mind is busy generating automatic, negative thoughts about yourself, your future, your world, or all these.

3. Automatic, negative thoughts are self-defeating. They cause painful, distressful feelings with associated behaviors and physical responses.

4. To feel better, you need to learn about the types of thoughts generated by your emotional mind and identify the emotional cycle in which you are trapped.

5. It is not uncommon to feel a combination of negative emotions. This means that you are trapped in more than one negative cycle and that your emotional mind is very busy generating automatic, negative thoughts.

6. Mapping out your own five roads provides an opportunity for you to study your emotional cycle(s). You will use the map of your own five roads in the next unit.

Unit 5 Assignment
Part A: Recognizing Negative Emotional Cycles

 Directions: Complete the following chart just as you did for Skill Session V. Next to each term, identify which of the five roads may apply, identify the negative emotional cycle that could apply, and then check whether you have experienced the term being described.

Term	Which of the Five Roads?	Possible Cycle? more than one may apply	Place a Check in This Column if You Have Experienced This
Rage	Feeling	Anger	✔
Sudden weight loss	physical response	depression	
I fail at everything.	thought	low self-esteem	
Being made fun of	trigger	anger	
Hopeless	feeling	depression	
Skipping school	behavior	anxiety	
You started this!	thought	anger	
No one will ever love me.	thought	depression	
Putting your fist through a wall	behavior	anger	
Clenched teeth	physical response	anxiety	
You're going to pay for this!	thought	anger	

Unit 5 Assignment-Part B: Mapping Out Your Own Emotional Cycle

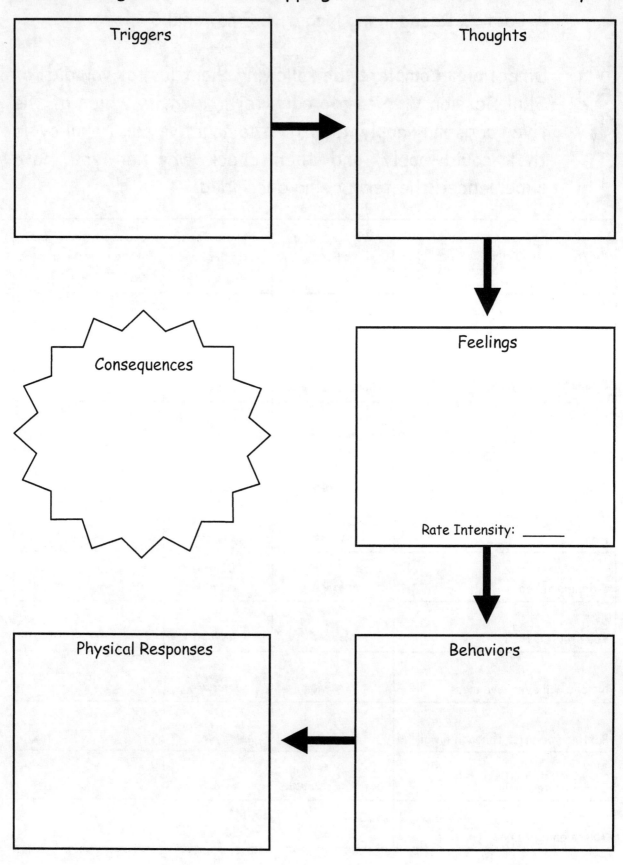

Mapping Out Your Own Emotional Cycle

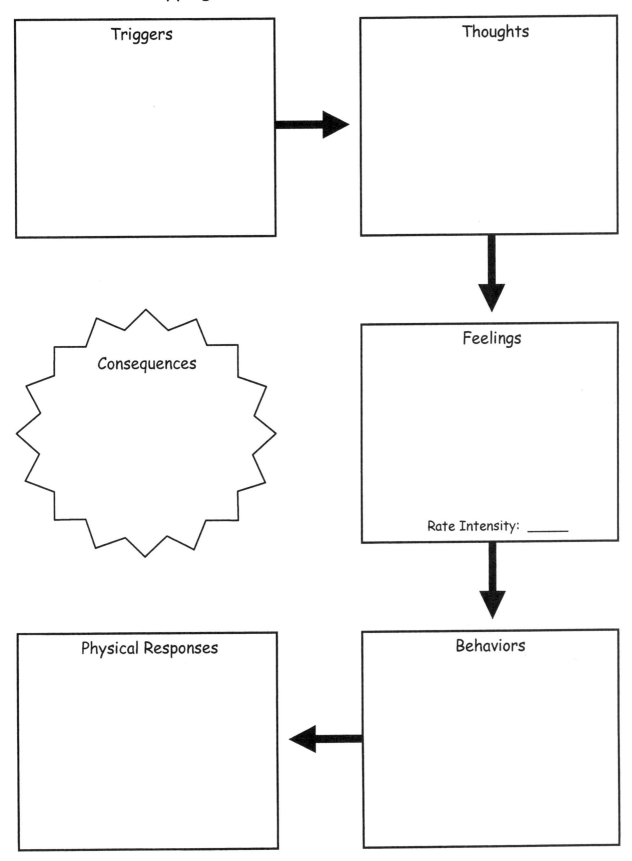

Mapping Out Your Own Emotional Cycle

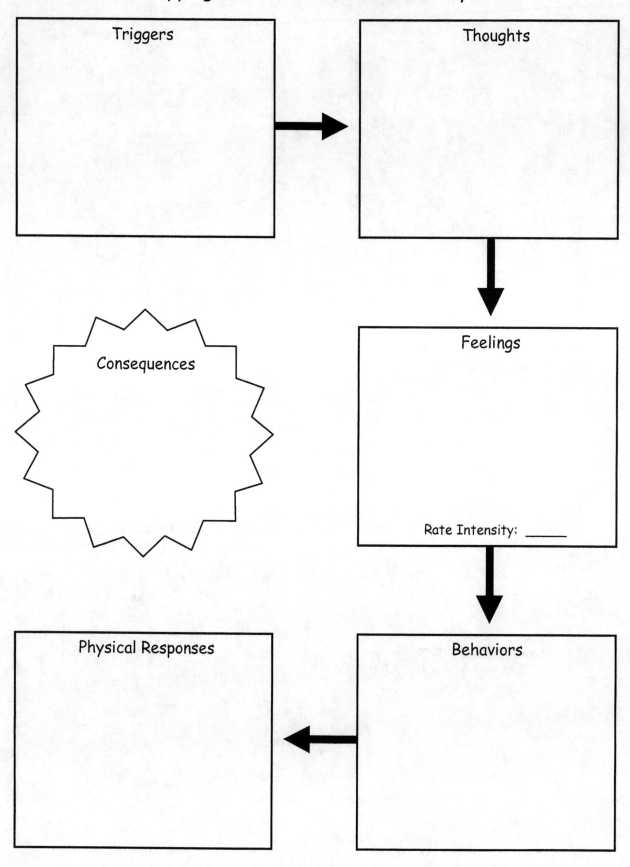

Triggers

Thoughts

Consequences

Feelings

Rate Intensity: _____

Physical Responses

Behaviors

Mapping Out Your Own Emotional Cycle

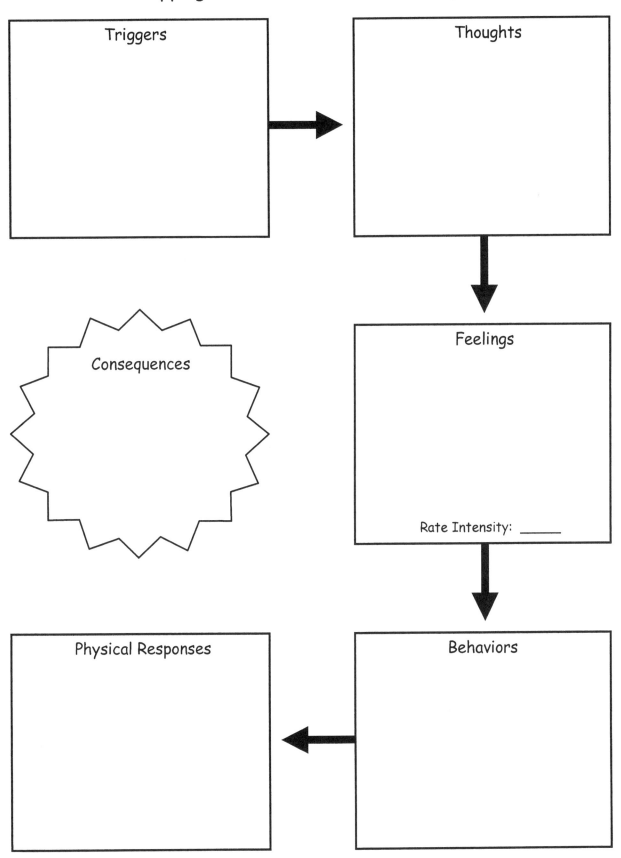

Triggers

Thoughts

Consequences

Feelings

Rate Intensity: _____

Physical Responses

Behaviors

Mapping Out Your Own Emotional Cycle

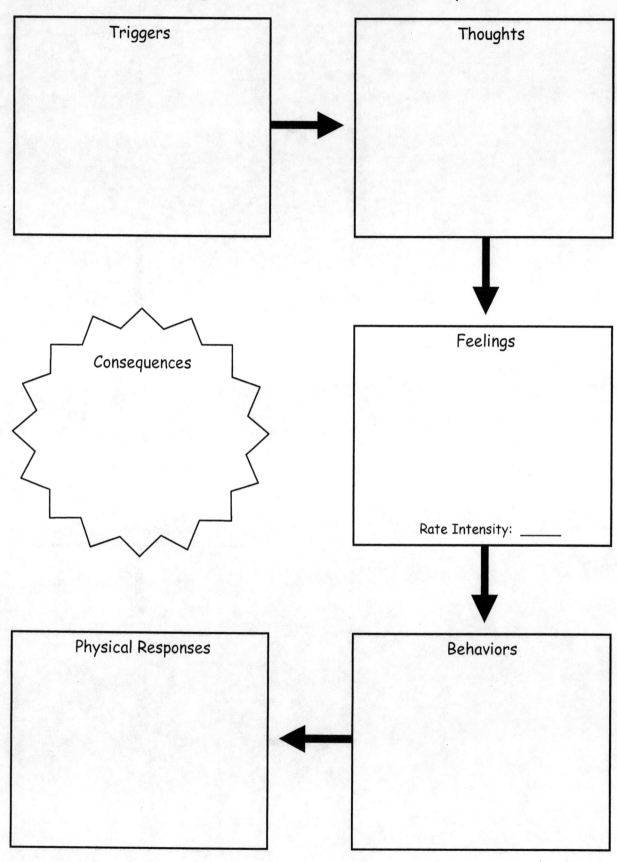

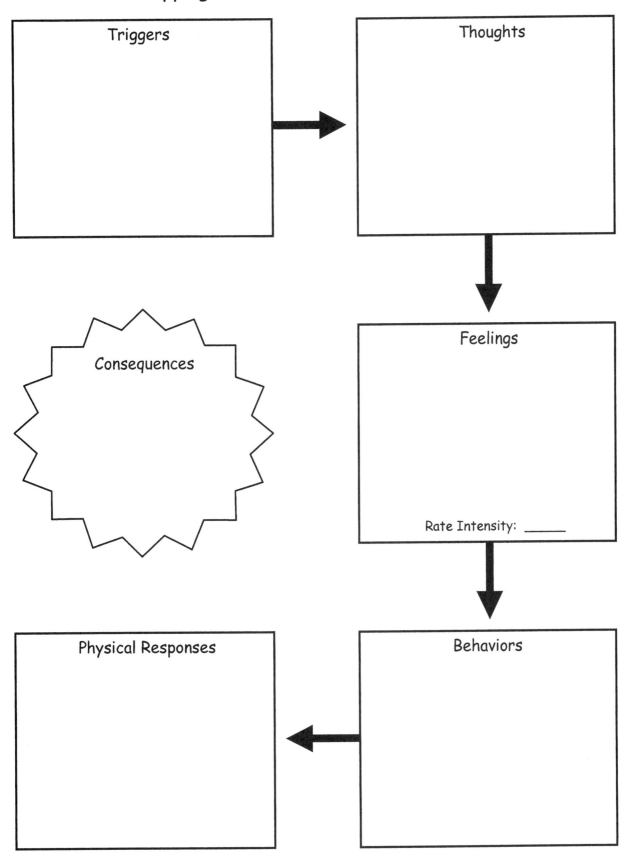

Mapping Out Your Own Emotional Cycle

Triggers

Thoughts

Consequences

Feelings

Rate Intensity: _____

Physical Responses

Behaviors

Before We Move On

Review:
Unit 5: The Cyclones of the Emotional Mind

1. Let's take a few minutes to review the key points on page 56.

2. What questions, if any, do you have about Unit 5?

3. Let's go over the weekly assignment on pages 57 and 58.

4. Mapping out your emotional cycle(s) is a very important skill in Mood Management. Let's take this opportunity to put several "maps" on the flip chart and discuss them. Who would like to volunteer to share his or her map?

Preview:
Unit 6: The Battle Lines Are Drawn

1. You've had some practice mapping out your own emotional cycles and listening carefully for the automatic, negative thoughts of your emotional mind. Let's take a closer look at your maps.

   a. Where is the emotional mind depicted on the map?
   b. Where do you think we should create a detour in the map to "break free" from the emotional traffic jam?
   c. In Unit 6, you will learn exactly where to take this detour.

Unit 6: The Battle Lines Are Drawn

The Wellness Mind Versus the Emotional Mind...

Helpful Hint
As we read this section out loud, <u>underline</u> or **highlight** *things that "hit home" for you!*

Now that you have learned the basic skills of Mood Management, it's time to move on to some of the more advanced skills found in your Mood Management toolbox. These advanced skills will require that you become much more active during the next several weeks. Not only will you be practicing these skills in our group meetings and as weekly assignments but also you will be using them each time you experience a strong, negative feeling. As you know, strong, negative feelings are a sign that you're stuck in an emotional traffic jam, and to get your traffic moving you need to practice these advanced skills over and over. Thus, you will be using them repeatedly in the following weeks.

You will recall from the previous unit that cycles of anger, depression, anxiety, and low self-esteem are perpetuated by the emotional mind's interpretation of a trigger. This interpretation includes automatic, negative thoughts about yourself, how you view your life, and how you envision the future. These thoughts just seem to "pop" into your mind for no reason. As a matter of fact, because these thoughts happen so often and so automatically they often go unnoticed. This means it is quite probable that you are not aware of the thoughts as they are happening. What you are aware of, however, is the negative feeling associated with the automatic thought. Thus, strong, negative feelings serve as a warning that your emotional mind is up and running, generating automatic, negative thoughts that keep you stuck in your emotional traffic jam.

To get your traffic moving and free yourself from the negative emotional cycle that has you trapped, you must learn to turn up the volume of your wellness mind and listen carefully to the realistic, positive thoughts that it is capable of generating. This advanced skill is the key that can unlock the door of the negative emotional cycle that has been problematic for you. Once unlocked, you can open the door, step out of the negative cycle, and view your world from a new perspective.

Viewing your world from a new perspective requires that your wellness mind becomes more powerful than your emotional mind. This is accomplished by turning down the volume of your emotional mind and listening more carefully to the thoughts generated by your wellness mind. Although these realistic, positive thoughts will serve to guide you out of your negative emotional cycle, they are often overpowered by the automatic, negative thoughts of the emotional mind. Thus, before you can get your wellness mind up and running, you will have to first put your emotional mind in neutral.

Putting your emotional mind in neutral requires that you become a master at recognizing the negative, self-defeating thoughts that it generates. Your ability to recognize these thoughts is a critical component of your journey toward emotional wellness. To turn down the volume of your emotional mind, you actually have to challenge the self-defeating thoughts that have caused you such difficulty. You cannot challenge your self-defeating thoughts unless you have first learned how to recognize them.

Before we introduce the skill required to challenge the self-defeating thoughts from your emotional mind, let's take a few minutes to review some of the information you learned in previous units about these automatic, negative thoughts. The more you know about these thoughts, the better equipped you will be to challenge them, which in turn will enable you to turn on your wellness mind.

One of the key features of self-defeating thoughts is that they often contain terms such as "should," "never," "always," "all the time," "everything," and "if/then." These terms are problematic because they convince you that there is no possibility of change. When these terms are operating, you view things in your life as either good or bad, right or wrong, etc., which leaves no room for gray. Since your wellness mind operates "in the gray," however, it is important to pay attention to these "all-or-nothing" terms that are generated by your emotional mind.

Another key feature of self-defeating thoughts is that they represent what you believe about yourself, your future, and your world. Thus, they tell you something about your core beliefs. What do you believe about yourself? What do you believe about your future and your world? What do you believe about other people? The list could go on and on, but the important point here is to remember that your self-defeating thoughts represent a core belief about how you experience your life.

For example, if you believe deep down inside your inner core that you are inadequate, your self-defeating thoughts will reflect this belief. You are likely to have thoughts such as "I can never do anything right," "This is too hard for me," and "No matter how hard I try, I will never get it." Thus, the core belief is "I am inadequate," and the self-defeating thoughts reflect this. In other words, your self-defeating thoughts are actually clues to your beliefs. Let's take a look at another example of how self-defeating thoughts are clues to your core beliefs. Suppose that deep down inside your inner core you believe that people cannot be trusted. In this belief system, you are likely to have thoughts such as "If I confide in him, then he'll just end up hurting me," "People always let me down," and "I really prefer to be by myself anyway." Again, the self-defeating thoughts provide a clue to the beliefs you hold about yourself, your future, and your world.

Another feature of self-defeating thoughts is that they are automatic. They come on fast and furious and cause an emotional storm that is so strong that all your attention gets pulled to the negative feelings generated by the self-defeating thoughts. You get so caught up in the feeling that you get stuck in your emotional traffic jam and

can't seem to find your way out. Again, use this experience as a clue. When you feel overwhelmed by a strong emotion, remind yourself that your emotional mind is operating. When this happens, it's your job to turn down the volume of the automatic, negative thoughts you experience. You accomplish this by first recognizing and then challenging these self-defeating thoughts. Simply stated, your ability to successfully break free from your negative emotional cycle depends on your ability to outwit your emotional mind.

Time Out
for
Discussion

1. As we read this section, what things hit home for you? What did you underline or highlight?

2. The title of this unit suggests that it's time for you to challenge your emotional mind. What do you think about this?

3. Core beliefs develop over time because of the emotional mind's negative interpretation of triggers. This interpretation is often the same even when the triggers are different. Therefore, your emotional mind tells you the same thing over and over again in different situations. Over time you develop a set of core beliefs about yourself and how you view things.

   a. What are some of the core beliefs of teenagers?
   b. How are core beliefs different from the automatic, negative thoughts generated by the emotional mind?

Let's take a few minutes to discuss the list on page 70, and then we'll move on to challenge the emotional mind.

Characteristics of the Emotional Mind

- It generates automatic, self-defeating thoughts.

- It wants you to believe negative things about yourself, your future, and your world.

- It's fast and furious...0 to 60 in 8.2 seconds.

- It likes to trick you.

- It uses key words such as "never," "should," "always," "if/then," and "everything."

- It keeps you stuck in your negative emotional cycle.

- It often gives you the same interpretation of different triggers that, over time, causes core beliefs to develop.

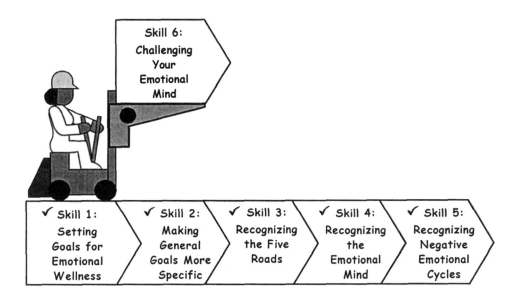

Now you're ready to learn how to challenge your emotional mind. This is a skill that requires keen awareness of the self-defeating thoughts it generates. When you become aware that your emotional mind is up and running, you need to ask yourself the question, "Where's the proof that these thoughts aren't true 100% of the time?" Challenging requires that you become your own detective and actively search for evidence that disproves the self-defeating thought.

Each time you use this skill, you actually "take on" your emotional mind and search for proof that discounts what it is trying to get you to believe about yourself, your future, and your world. How well you perform the process of challenging depends on how good a detective you are and how actively you search your environment for clues that disprove the self-defeating thoughts.

It's important for you to become familiar with the steps involved in the challenging process. Study the diagram on page 72, and use it as a guide as you complete your map to emotional wellness.

Challenging Your Emotional Mind

1.
Recognize Your Life Triggers
What difficult experiences have you had? What's in your trigger bank?

2.
Listen for Your Emotional Mind
Recognize the automatic, self-defeating thoughts generated by your emotional mind as it interprets your life triggers.

3.
Identify Your Core Beliefs
What is your emotional mind trying to get you to believe about yourself, your future, and your world?

4.
Identify Your Feelings
How do you feel in response to your automatic, self-defeating thoughts and core beliefs?

5.
STOP
Don't respond to your emotional mind.

6.
Look & Listen: Where's the Proof?
Scan your environment for evidence that disproves your self-defeating thoughts. How do you know they're not true 100% of the time?

7.
Re-Think
Replace your automatic, negative thoughts and core beliefs with realistic, positive thoughts from the wellness mind.

8.
Re-Examine Your Feelings
How do you feel in response to the thoughts generated by your wellness mind?

9.
Re-Act
Respond to the thoughts of your wellness mind. Make a commitment to change your "old ways."

Time Out for Discussion

1. At first, the challenging process may appear to be more complex than it actually is. Look closely at the diagram on page 72.

   a. What steps in the challenging process are you already familiar with?

b. Where have you seen these steps before?

2. Challenging your emotional mind enables you to get out of the emotional cycle that has had you trapped. It does this by creating a detour to the wellness mind. This detour takes place at an important point on your map. Study the following map:

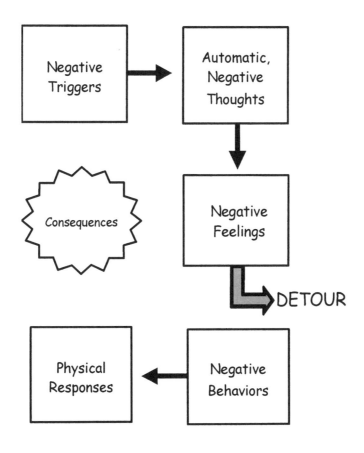

a. Why put the detour before you get to the box labeled "negative behaviors"?

b. If you change your behaviors, what is likely to happen to the consequences you experience?

3. Let's take a look at the entire map to the wellness mind. Notice that the boxes on the map are the same as the steps of the Challenging process.

a. Let's take a few minutes to go over the map and plot out the route to the wellness mind.

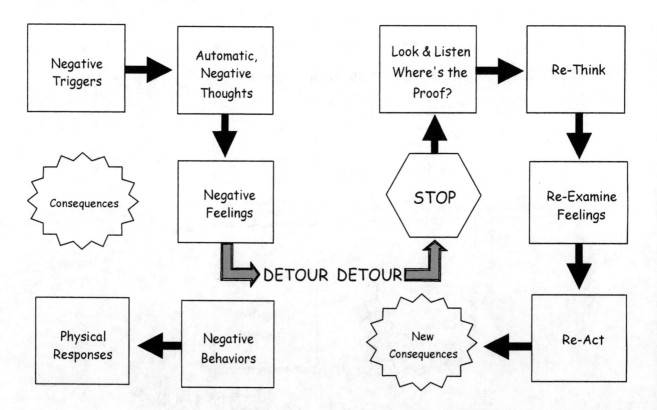

b. Which side of the map is controlled by the emotional mind?

c. Which side of the map is controlled by the wellness mind?

Stop Here...
Complete Skill Session VI:
Challenging Your Emotional Mind

Directions: The skill session for this unit contains a Challenging Map found on pages 76 and 77. Use the map to challenge the negative, self-defeating thoughts of your emotional mind. To start the map, fill in the left side using a previous example from Unit 5 or choose a different situation that you haven't mapped out yet.

1. We will work on the Challenging Maps together. Who would like to volunteer to have their map put on the flip chart and discussed?

2. What happens to the intensity of your feelings once you challenge the self-defeating thoughts from the emotional mind?

3. When would it be useful to use the Challenging Map? We will discuss this more in Unit 7.

Challenging Map - Side A

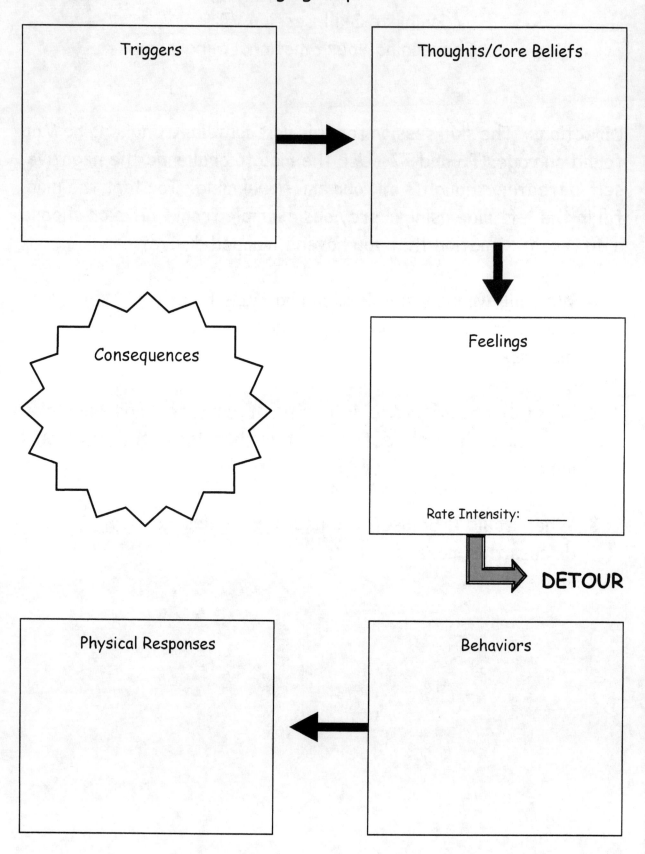

Triggers

Thoughts/Core Beliefs

Consequences

Feelings

Rate Intensity: _____

DETOUR

Physical Responses

Behaviors

Challenging Map - Side B

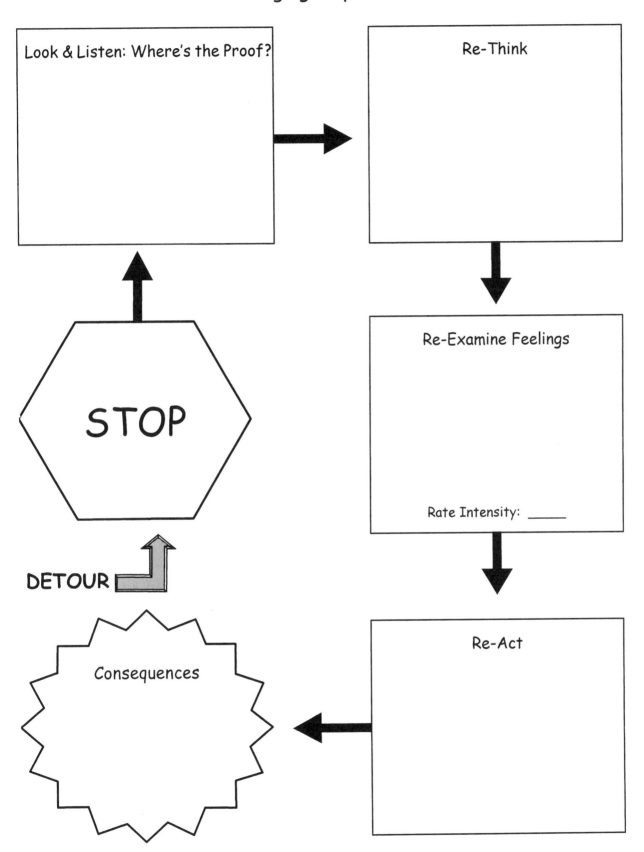

Look & Listen: Where's the Proof?

Re-Think

STOP

Re-Examine Feelings

Rate Intensity: _____

DETOUR

Consequences

Re-Act

Key Points to Remember...

Unit 6: The Battle Lines Are Drawn

1. Strong, negative emotions are clues to your emotional mind. When you feel caught in your anger, depression, anxiety, or low self-esteem, it is because your emotional mind is busy generating automatic, self-defeating thoughts.

2. Getting out of your negative emotional cycle requires that you "turn down" the volume of your emotional mind.

3. Challenging is the skill used to "turn down" the volume of your emotional mind.

4. Challenging is an advanced skill that requires constant practice. Use the Mood Management Challenging Map as a tool to help you confront your emotional mind.

5. Challenging your emotional mind results in new behaviors. These new behaviors will have different consequences than the ones you experienced while trapped in your emotional cycle.

6. Challenging is a skill that can be used to help you confront both automatic, negative thoughts and negative core beliefs.

7. Use your Challenging Map as a tool each time you become aware that your emotional mind is trying to get the better of you.

Unit 6 Assignment: Challenging

Challenging Maps are found on pages 80-85. Complete these maps as you encounter triggers that set your emotional mind off. If you need additional Challenging Maps, you can use the ones in Appendix A.

Challenging Map

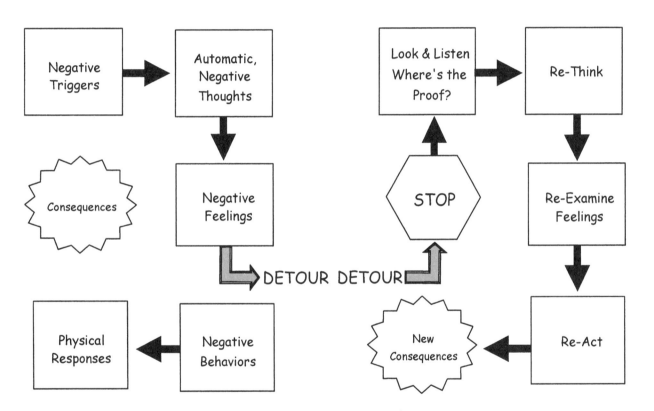

Challenging Map - Side A

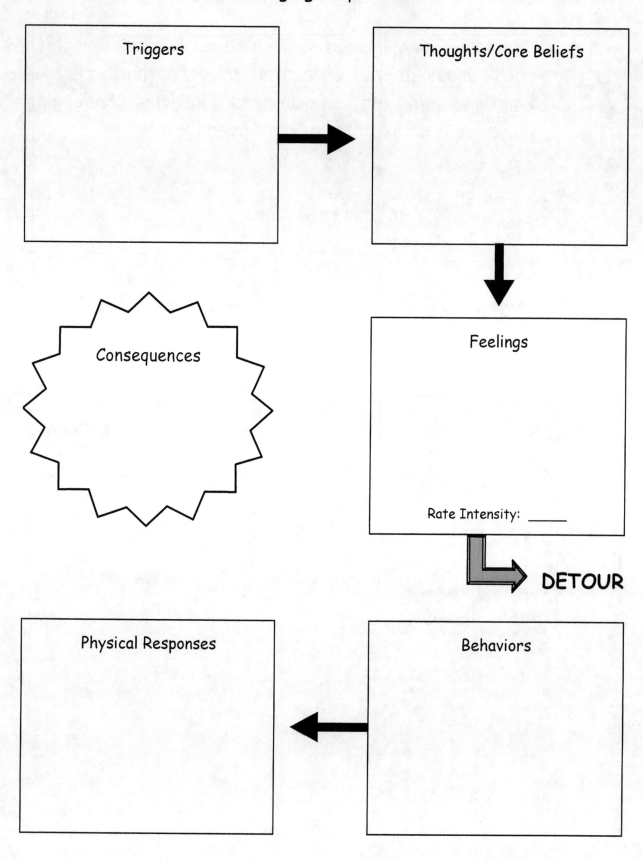

Triggers

Thoughts/Core Beliefs

Consequences

Feelings

Rate Intensity: _____

DETOUR

Physical Responses

Behaviors

Challenging Map - Side B

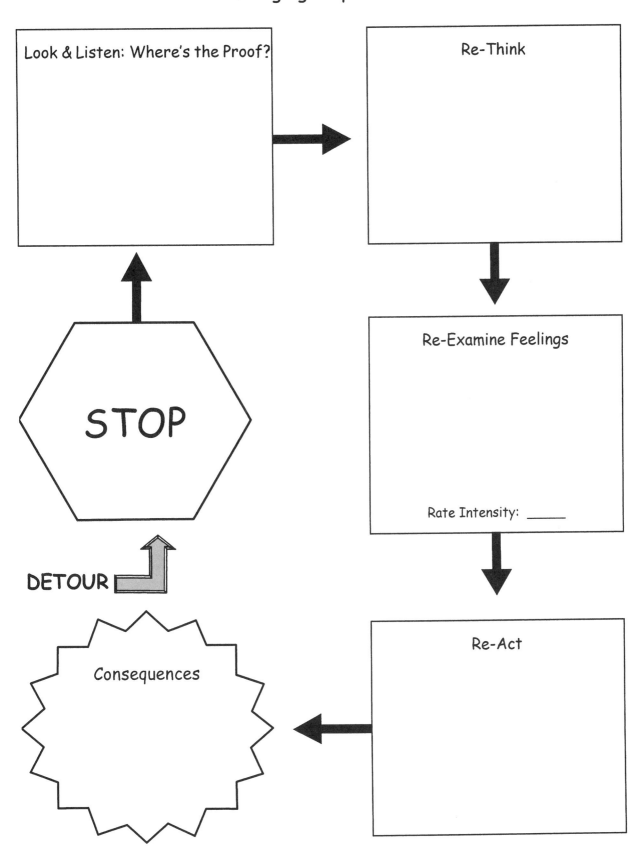

Look & Listen: Where's the Proof?

Re-Think

STOP

Re-Examine Feelings

Rate Intensity: _____

DETOUR

Consequences

Re-Act

Challenging Map - Side A

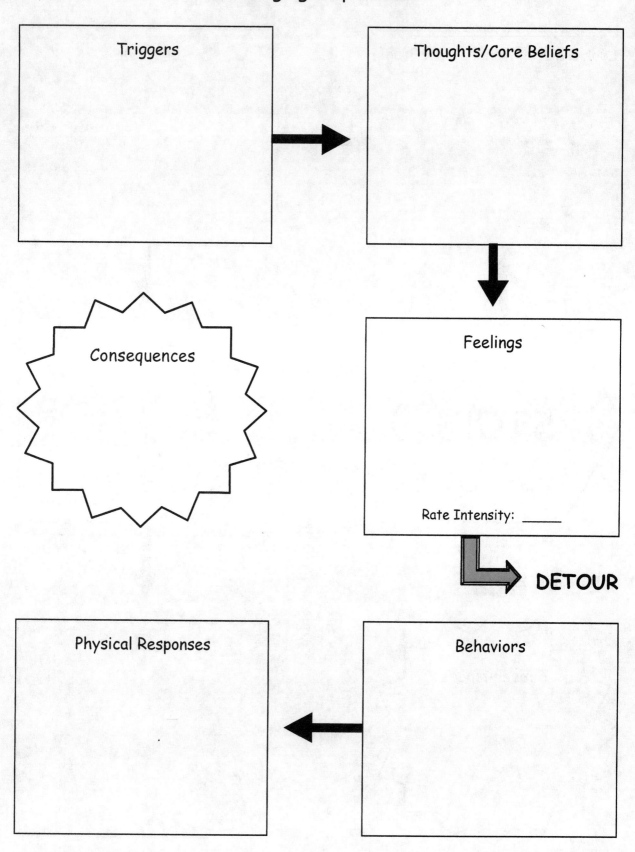

Triggers

Thoughts/Core Beliefs

Consequences

Feelings

Rate Intensity: _____

DETOUR

Physical Responses

Behaviors

Challenging Map - Side B

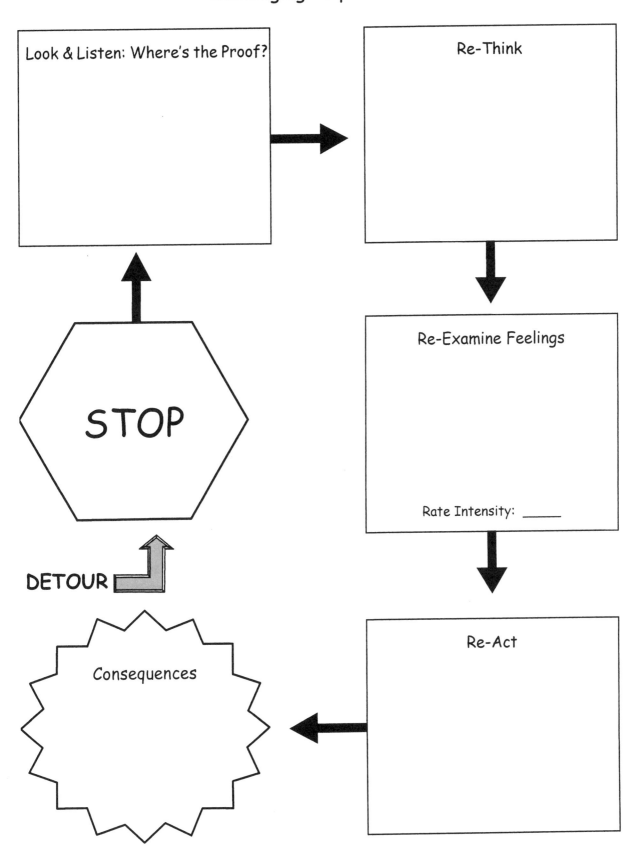

Look & Listen: Where's the Proof?

Re-Think

STOP

Re-Examine Feelings

Rate Intensity: _____

DETOUR

Consequences

Re-Act

Challenging Map - Side A

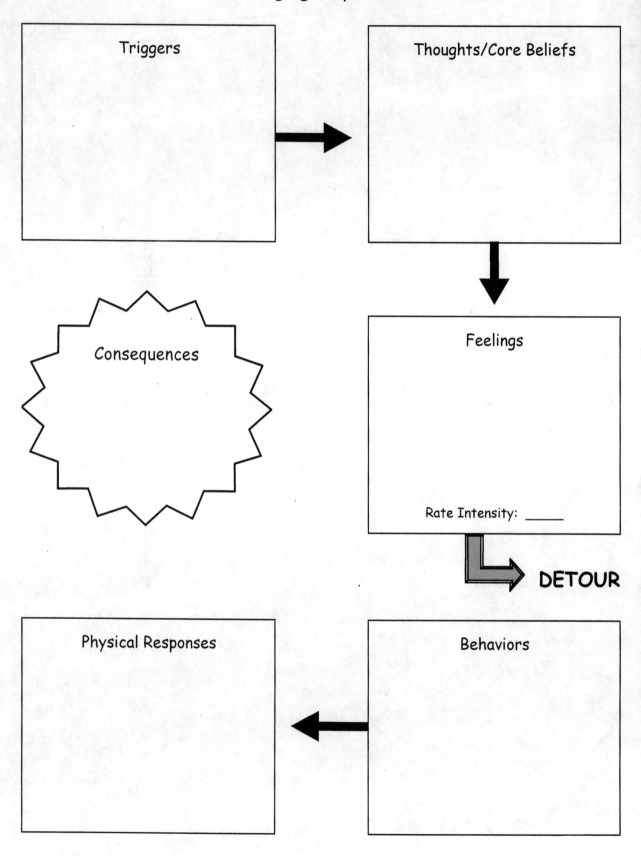

Triggers

Thoughts/Core Beliefs

Consequences

Feelings

Rate Intensity: _____

DETOUR

Physical Responses

Behaviors

Challenging Map - Side B

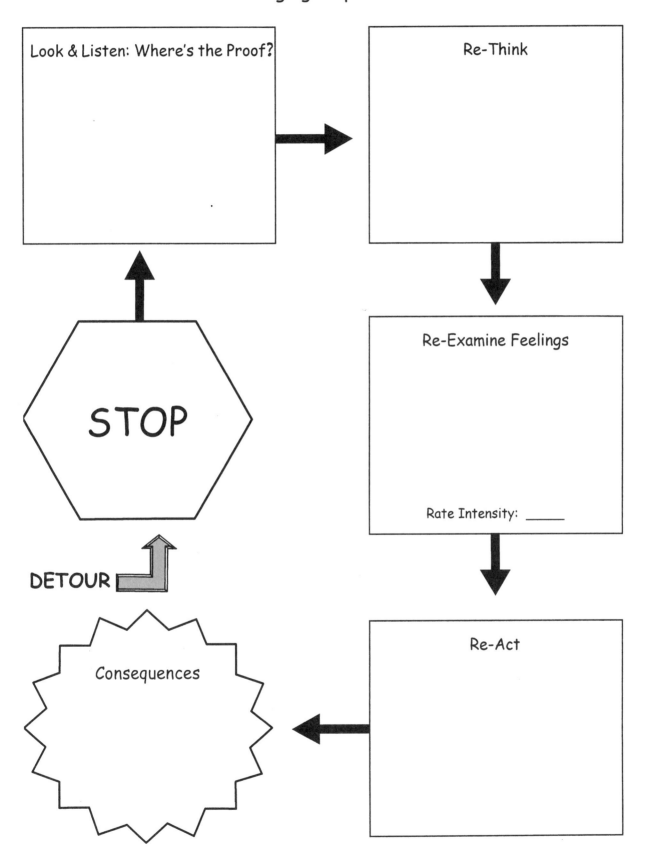

Look & Listen: Where's the Proof?

Re-Think

Re-Examine Feelings

Rate Intensity: _____

STOP

DETOUR

Consequences

Re-Act

Before We Move On

```
┌──────────────────────────────────────────┐
│                Review:                     │
│   Unit 6: The Battle Lines Are Drawn       │
└──────────────────────────────────────────┘
```

1. Let's take a few minutes to review the key points on page 78.

2. What questions, if any, do you have about Unit 6.

3. Let's go over the weekly assignment by reviewing someone's Challenging Map. Who would like to volunteer?

4. Let's take this opportunity to put several other Challenging Maps on the flip chart and discuss them. Who would like to volunteer?

5. Let's go all the way back to page 24. Look at the number that you put in the center of the traffic jam. Reassign a number based on how you feel now. Has anyone's number changed? Let's discuss this.

```
┌──────────────────────────────────────────┐
│                Preview:                    │
│   Unit 7: Lights, Camera...Action!!        │
└──────────────────────────────────────────┘
```

1. Challenging your emotional mind is a key Mood Management skill. Although it is very important to continue to practice challenging, there is another skill that you need to learn.

2. What do you think the title of this unit might mean?

Unit 7:
Lights, Camera...
Action!!

Helpful Hint
As we read this unit out loud, underline or highlight things that "hit home" for you!

Throughout this Mood Management Program, you have been introduced to various skills designed to help you better manage the negative feelings that have kept you stuck in an emotional traffic jam. As you now know, effectively managing these emotions depends on your ability to recognize the self-defeating thoughts generated by your emotional mind. This, in turn, allows you to get your wellness mind up and running through the use of the Mood Management skill called challenging.

Once you have mastered this skill, you will be able to call on your wellness mind each time you encounter an emotional bump in the road during your continued journey toward emotional wellness. Remember, however, that challenging the self-defeating thoughts generated by your emotional mind is a skill that takes continual practice. The more you practice this skill, the better you will become at managing the difficult emotions you experience in response to the automatic, negative thoughts generated by your emotional mind. This, in turn, will enable you to eventually reach the goals for emotional wellness that you set for yourself at the beginning of the Mood Management Program. You set these goals and now have the skills to reach them. It is at this point, however, that you must make a choice. That's right. You must choose between emotional wellness and remaining stuck in your emotional traffic jam. Emotional wellness requires a continued commitment to practicing the skills that you have learned, which in turn takes hard work and much effort. Although emotional wellness is certainly worth your time and effort, the decision to continue to "go for it" is ultimately up to you.

At times, it may seem easier to remain stuck in your emotional traffic jam. After all, you have probably been there for quite some time, and the pull to remain there can be very strong. Despite this pull, however, we encourage you to "take the bull by the horns" and continue on your journey to emotional wellness. The rewards of becoming emotionally well are many, and they cannot be understated. Remember, emotionally well adolescents mind their IMAGE.

Since challenging is so important to achieving emotional wellness, let's review this skill one more time. Remember, the first step in challenging is to identify your life triggers, both past and present, and then recognize the automatic, negative thoughts your emotional mind generates in response to the triggers. These thoughts, as shown in this diagram, are about you, your world, your future, or all three.

Once you recognize these thoughts, your related core beliefs, and the negative emotions they cause, you *stop*, *look*, and *listen* to evidence from your environment that shows that these automatic, negative thoughts are not true 100% of the time. Scanning your environment for this evidence takes a critical eye and good hearing. It is likely that this evidence has always been in your environment, but you have probably overlooked it while stuck in your emotional traffic jam. Emotional traffic jams tend to dull our senses, so it is critical to pay very close attention to the evidence that exists in your environment that disproves the automatic, negative thoughts you have about yourself, your world, your future, or all three.

Once you have collected your evidence, you use it to generate more positive, realistic thoughts from your wellness mind. Remember, Mood Management is not simply the "power of positive thinking." Mood Management requires the ability to take a realistic look at yourself, your world, and your future, which is accomplished by challenging your emotional mind and turning on your wellness mind.

The final steps in challenging include reevaluating the negative emotions caused by the self-defeating thoughts of the emotional mind and learning to react differently to life's triggers. If you have challenged your self-defeating thoughts effectively, your emotions will be more positive. This means that you will be able to overcome the overwhelming feelings of anger, depression, anxiety, and low self-esteem that have caused you such difficulty. This is the power of Mood Management. Through the use of challenging, your thoughts and feelings become more positive, which in turn frees you from your emotional traffic jam.

After you have completed many Challenging Maps and understand the process of challenging, you may prefer to use a thought record as the tool you use to challenge your emotional mind. Thought records have the same steps as your Challenging Maps. Of course, if you prefer to use the Challenging Maps, go right ahead. The following is an example of a thought record. Additional thought records are found in Appendix C.

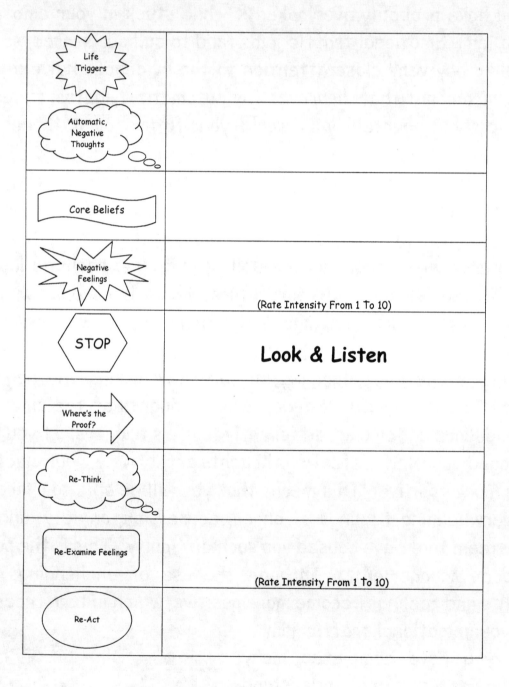

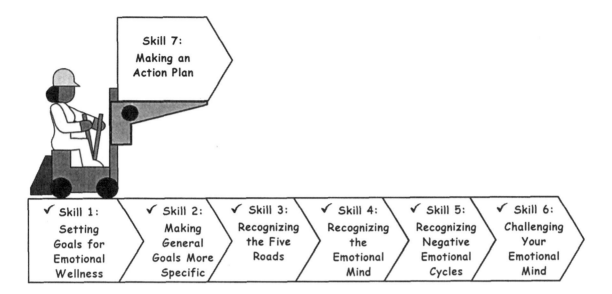

Although challenging is a very important Mood Management skill, it is important to note that "That's not all there is!" Challenging allows you to step out of your emotional traffic jam to experience your world in a more positive way. It's not enough to just step out of your traffic jam, however. After you step out, you have to keep your feet moving.

You do this in two ways. First, as we already stated, you continue to use the skill of challenging each time you hit an emotional bump in the road. The second thing you do to keep your feet moving is to make and follow an action plan. An action plan is an individualized plan that helps you get back on track and move toward the goals you set for yourself at the beginning of the Mood Management Program. To make an action plan, you need to ask yourself the following questions:

1. What emotion(s) has caused me difficulty?
2. What has my emotional difficulty stopped me from doing?
3. What small steps do I want to take to get going?

You must also remember the following rules about action plans:

1. Action plans are individualized plans! Your plan may look very different from someone else's plan.
2. Action plans require small increments. We want you to take very small steps toward your stated goals. Break down the goal into its smallest components and start from there.
3. Be patient with yourself. Your journey toward emotional wellness is a lifelong process. Plan carefully and have a safe trip!

Now, to write your action plan you will need to look back at the first two units in your *Mood Management Manual* to review the goals you set for yourself. Once you have identified one goal from your list that you want to achieve, fill in your action plan, make a commitment to put your plan into effect, and go for it!!

Let's take a closer look at what we mean by an action plan by reviewing the example on page 93. Remember, an action plan is an individualized plan designed to help you achieve the goals that you were having trouble accomplishing because you were stuck in an emotional traffic jam. Challenging helps get you out of the traffic jam, which in turn enables you to put your action plan into effect.

As you can see, action plans are designed to help you achieve the goals you set for yourself at the beginning of the Mood Management Program. As in the example on page 93, each of your overall goals probably has multiple smaller goals that you can begin to work on.

My Action Plan

The emotion(s) that has caused me difficulty is: _____anger_____

This emotion has stopped me from: __getting along with others__

The small goal I would like to achieve is: getting along with one person

Goal(s)	Specific Steps to Take	Evaluate Results
Get along better with my boyfriend/girlfriend.	1. Listen carefully to what he/she says. 2. Buy him/her a greeting card just because. 3. When I feel angry, count to 10 before saying anything. 4. When I feel angry, tell him/her that I need a time-out. 5. When I feel angry, fill out a Challenging Map.	1. No problems. 2. He/she was surprised. 3. I had to count to 30.
Get along better with my parents.	1. Do a chore without being asked. 2. Invite my parents out to breakfast. 3. When I feel angry, take a time-out. 4. When I feel angry, fill out a Challenging Map.	1. They were surprised. 2. We had a nice time.
Get along better with my teachers.	1. Do my work. 2. Listen carefully.	

Time Out
for
Discussion

1. As we read this unit, what things hit home for you? What did you underline or highlight?

2. Action plans play a key role in the Challenging process. On the right-hand side of the Challenging Map, there's a box labeled Re-act. You can use your action plan to design a strategy that will help you react to difficult emotions in a positive way rather than falling back on the old behaviors that were part of your negative emotional cycle.

Let's brainstorm a list of specific steps or actions, besides those listed in column two on page 93, that can be used to react positively to the different negative emotional cycles. For example,

Negative Emotional Cycle	Old Behavior	Re-Act (specific step/action)
Depression	Withdraw	Exercise
Anger	Hitting things	Talk with someone

a. What other specific steps or actions can you think of to react to difficult emotions more positively?

b. How will reacting positively, rather than falling back on old behaviors, affect the consequences of your actions?

Stop Here...

Complete Skill Session VII:
Making an Action Plan

My Action Plan

The emotion(s) that has caused me difficulty is: _____

This emotion has stopped me from: _____

The small goal I would like to achieve is: _____

Goal(s)	Specific Steps to Take	Evaluate Results

 Key Points to Remember...

Unit 7: Lights, Camera...
Action!!

1. You have learned many skills that can enable you to manage your difficult emotions. It is now up to you to continue to practice these skills to achieve emotional wellness. It's your choice.

2. Challenging your emotional mind is a key Mood Management skill. Remember, recognize self-defeating thoughts; stop, look, and listen for evidence that disproves the thought; formulate a more realistic, positive thought; reevaluate your feeling and react more positively.

3. Action plans are individualized plans designed to help you achieve the goals you set for yourself at the beginning of the Mood Management Program. Action plans require small steps. Plan carefully, and set yours in motion.

Unit 7 Assignment: Action Plans

Complete a Mood Management action plan for each goal you stated in Unit 1. If you need additional action plans, you can use the ones in Appendix B.

My Action Plan

The emotion(s) that has caused me difficulty is: _____

This emotion has stopped me from: _____

The small goal I would like to achieve is: _____

Goal(s)	Specific Steps to Take	Evaluate Results

My Action Plan

The emotion(s) that has caused me difficulty is: _____

This emotion has stopped me from: _____

The small goal I would like to achieve is: _____

Goal(s)	Specific Steps to Take	Evaluate Results

Appendix A

Challenging Maps

Challenging Map - Side A

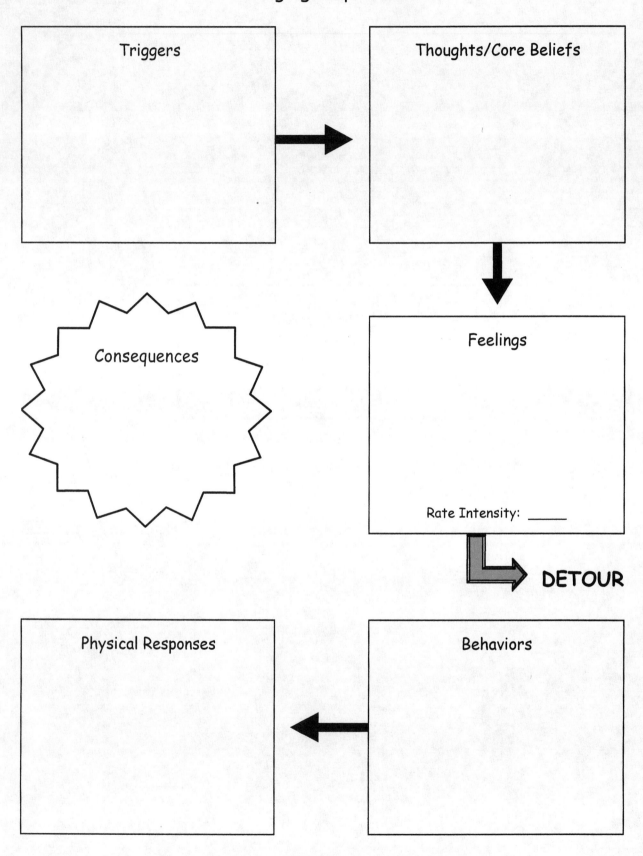

Challenging Map - Side B

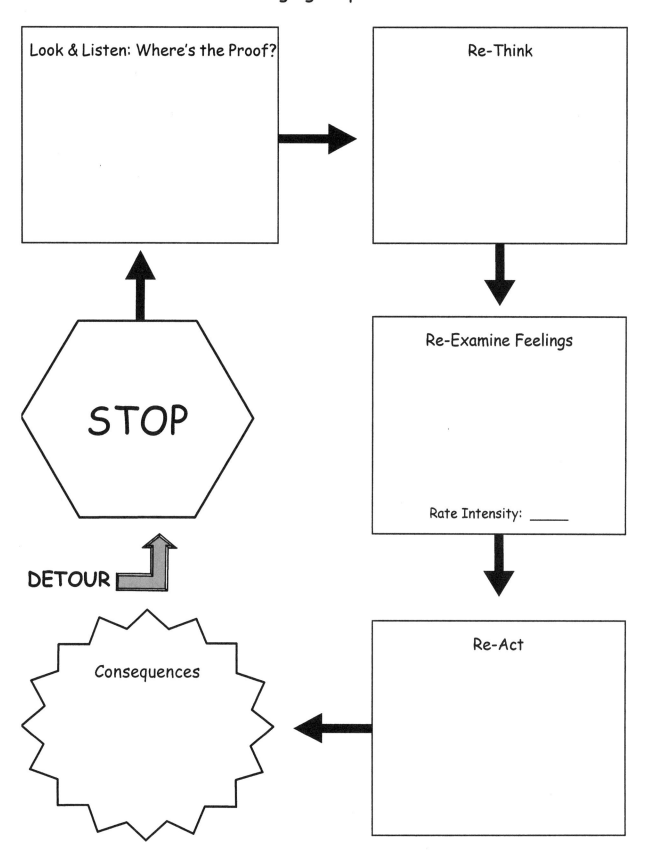

Challenging Map - Side A

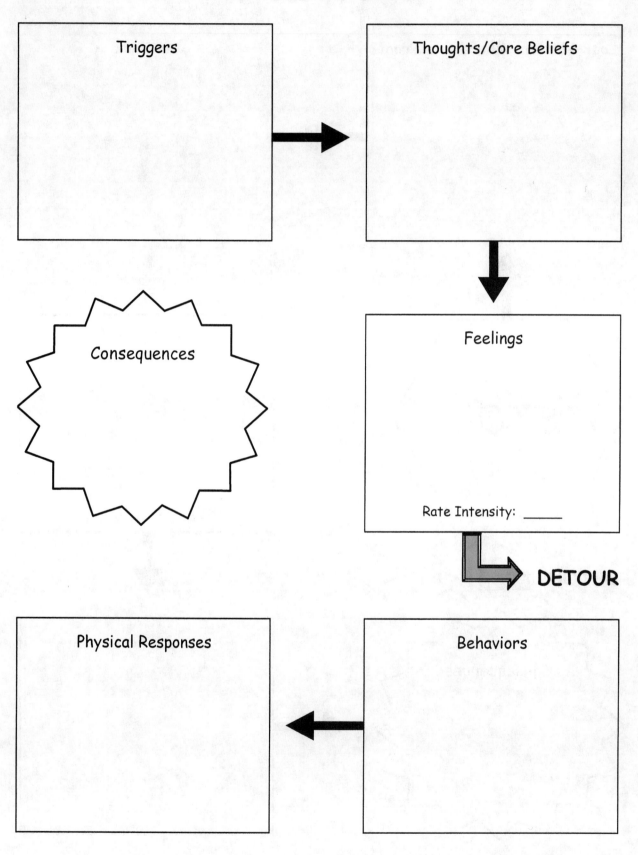

102

Challenging Map - Side B

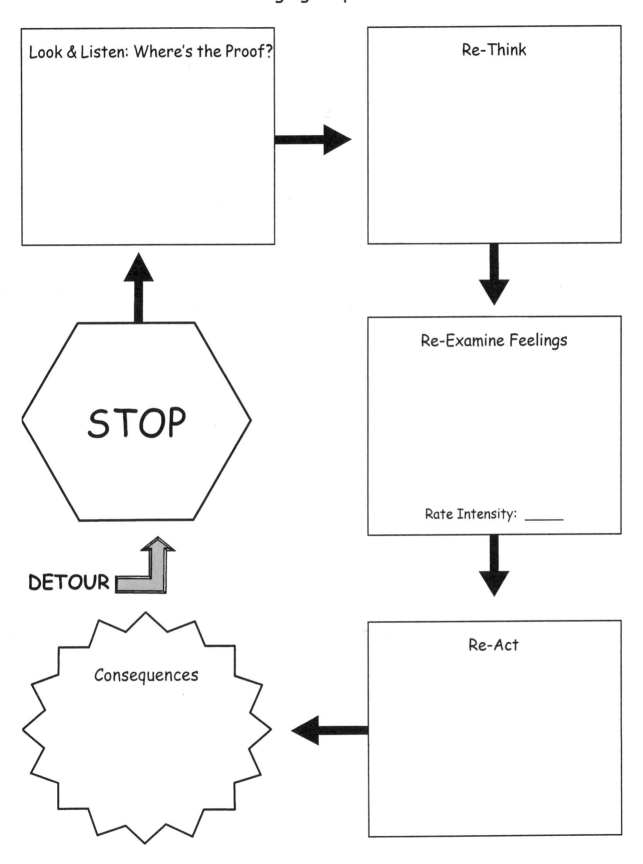

Look & Listen: Where's the Proof?

Re-Think

STOP

Re-Examine Feelings

Rate Intensity: _____

DETOUR

Consequences

Re-Act

Challenging Map - Side A

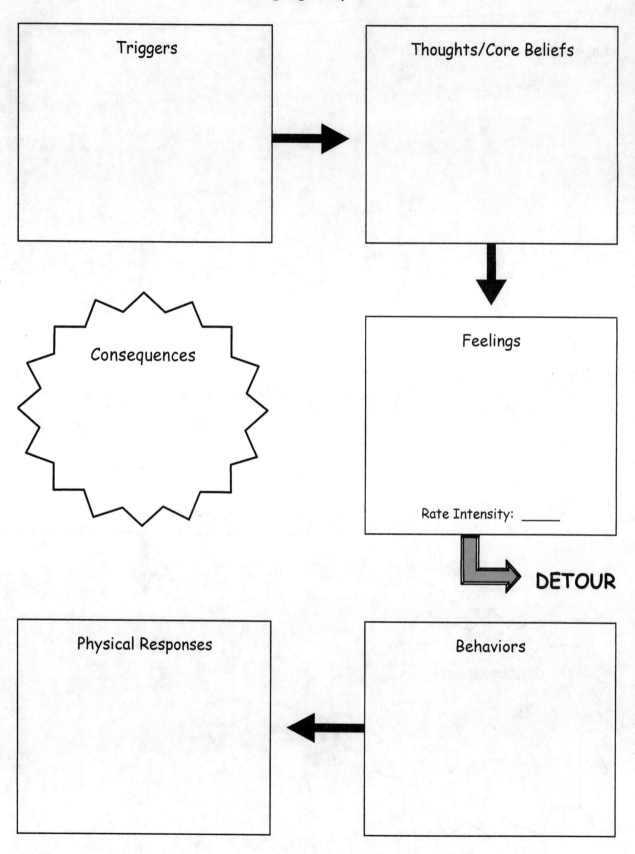

Triggers

Thoughts/Core Beliefs

Consequences

Feelings

Rate Intensity: _____

DETOUR

Physical Responses

Behaviors

Challenging Map - Side B

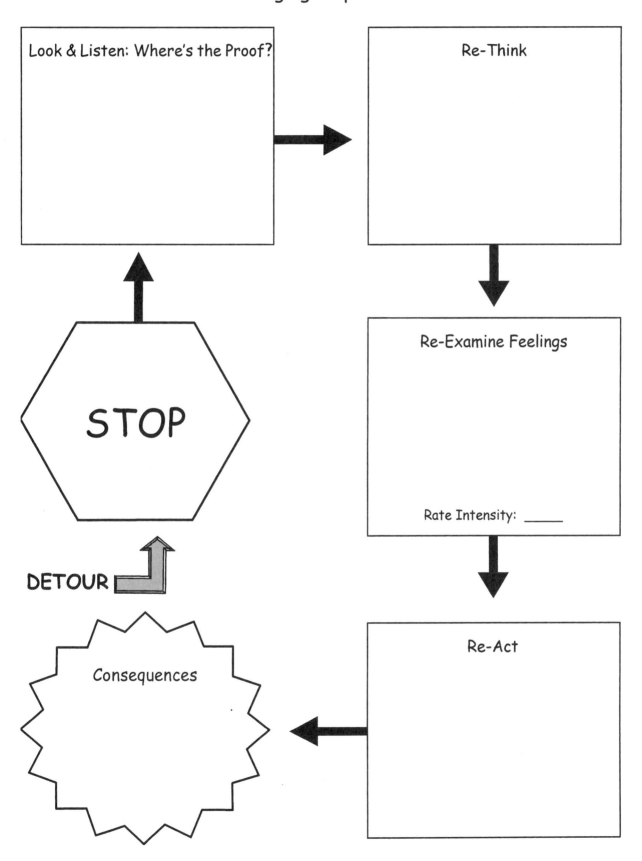

Challenging Map - Side A

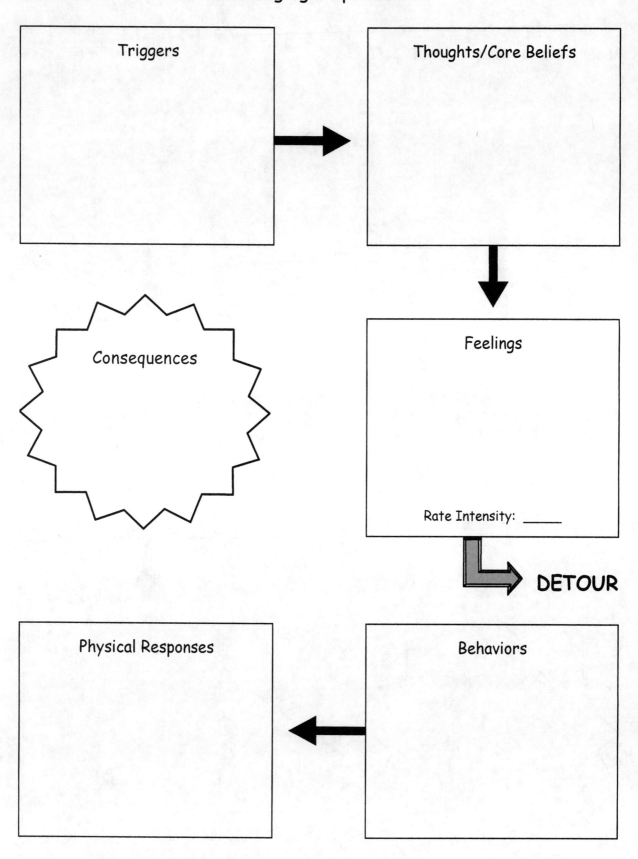

| Triggers | Thoughts/Core Beliefs |

Consequences

Feelings

Rate Intensity: _____

DETOUR

Physical Responses

Behaviors

Challenging Map - Side B

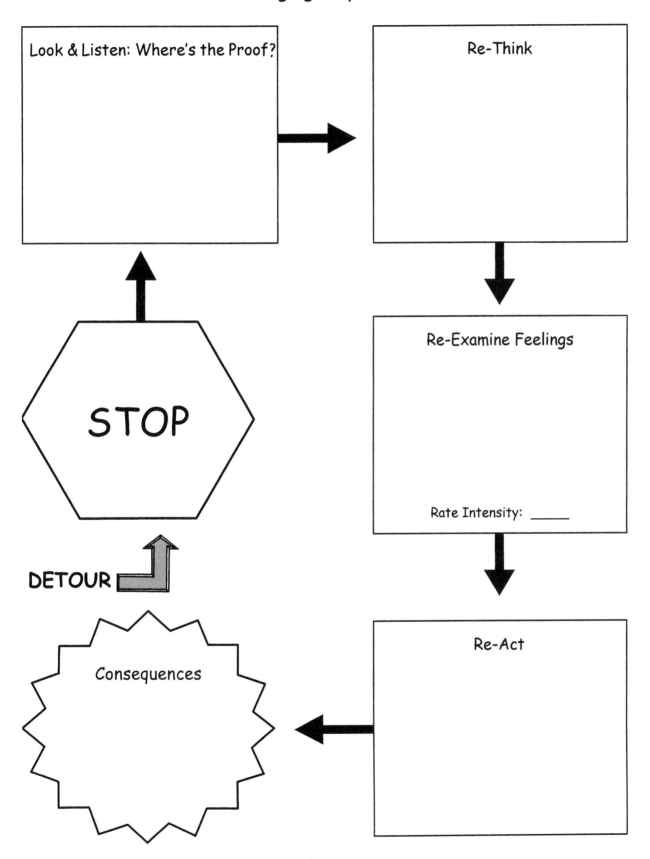

Challenging Map - Side A

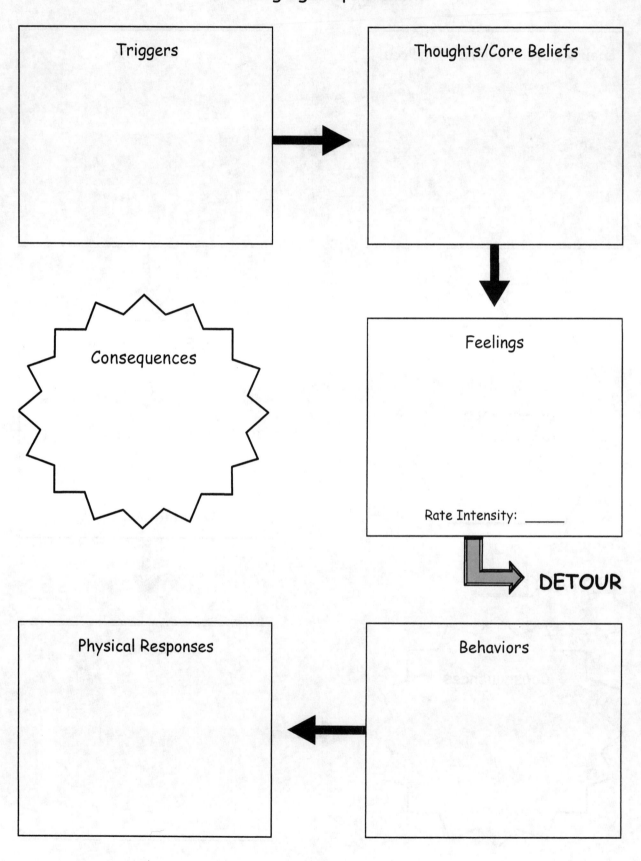

Challenging Map - Side B

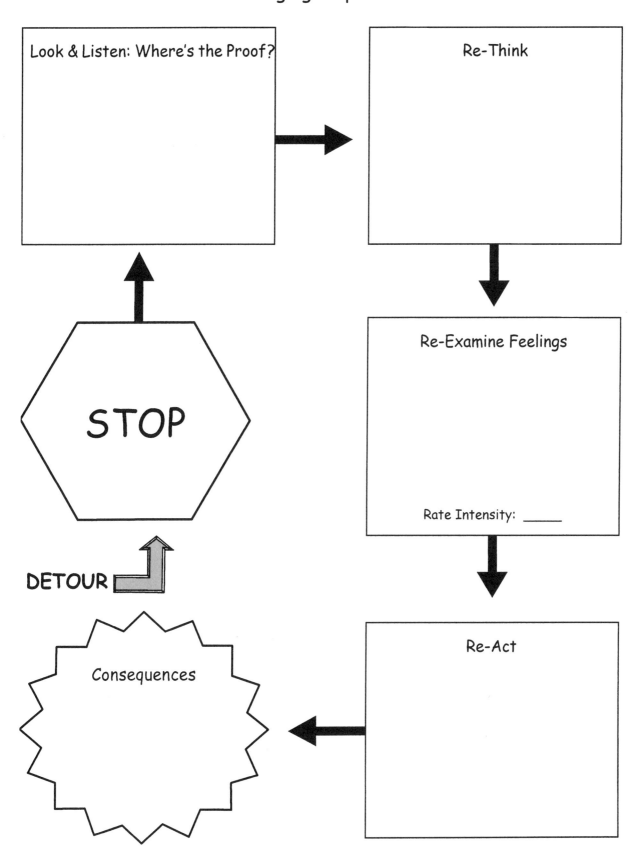

Challenging Map - Side A

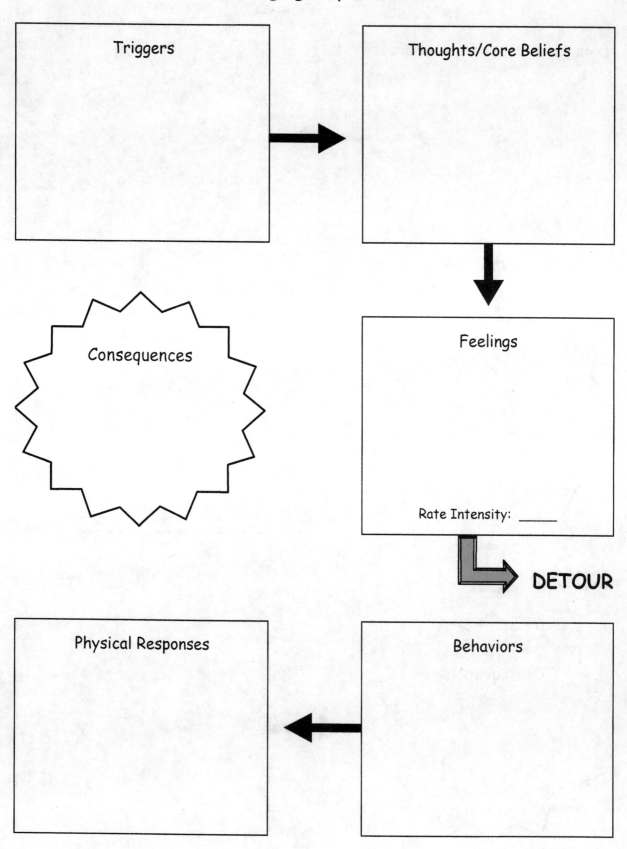

Triggers

Thoughts/Core Beliefs

Consequences

Feelings

Rate Intensity: _____

DETOUR

Physical Responses

Behaviors

Challenging Map - Side B

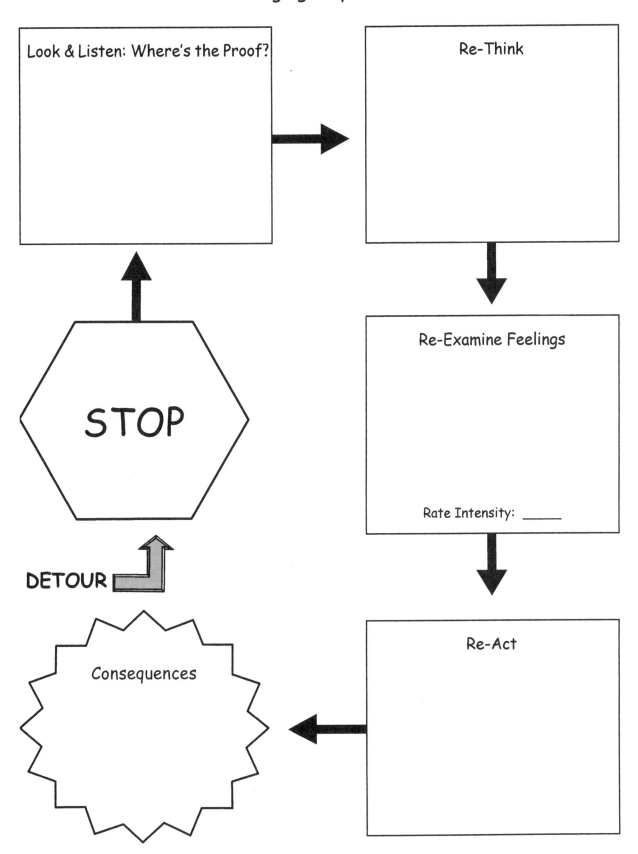

Look & Listen: Where's the Proof?

Re-Think

Re-Examine Feelings

Rate Intensity: _____

STOP

DETOUR

Consequences

Re-Act

Challenging Map - Side A

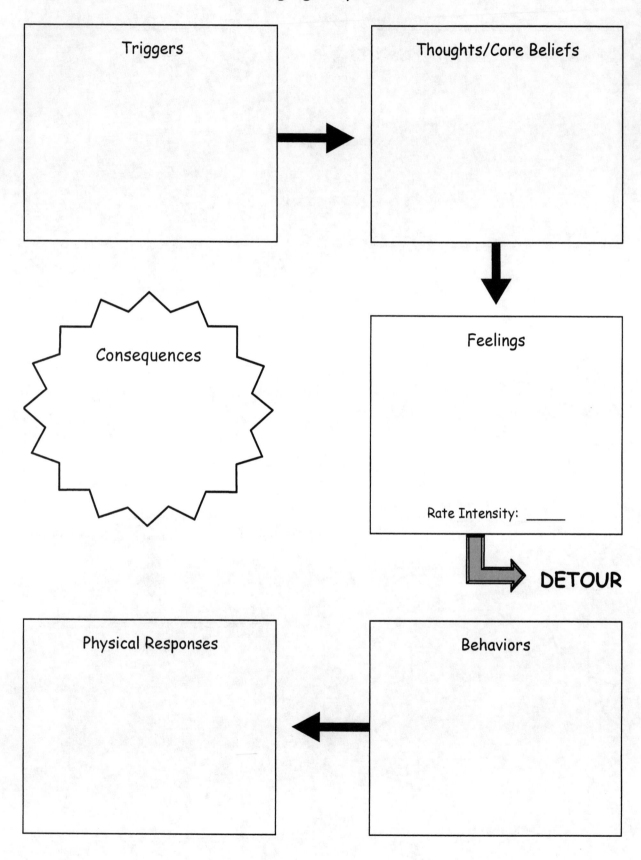

Challenging Map - Side B

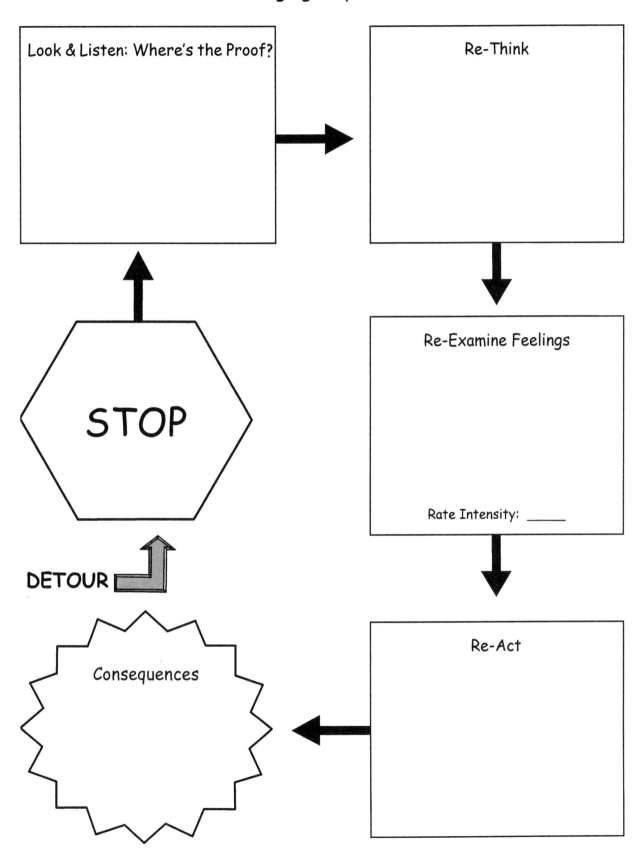

Look & Listen: Where's the Proof?	Re-Think

STOP

DETOUR

Re-Examine Feelings

Rate Intensity: _____

Consequences

Re-Act

Challenging Map - Side A

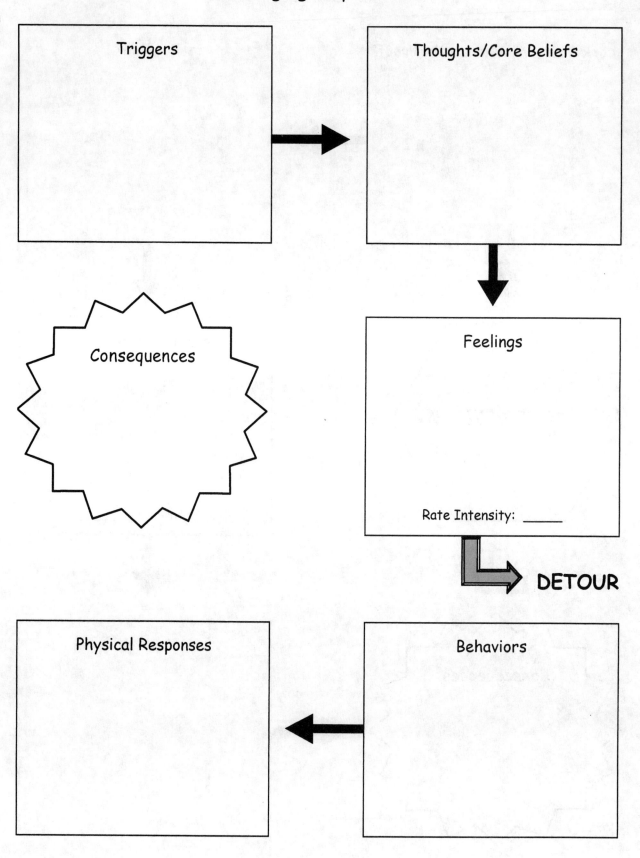

Triggers

Thoughts/Core Beliefs

Consequences

Feelings

Rate Intensity: _____

DETOUR

Physical Responses

Behaviors

Challenging Map - Side B

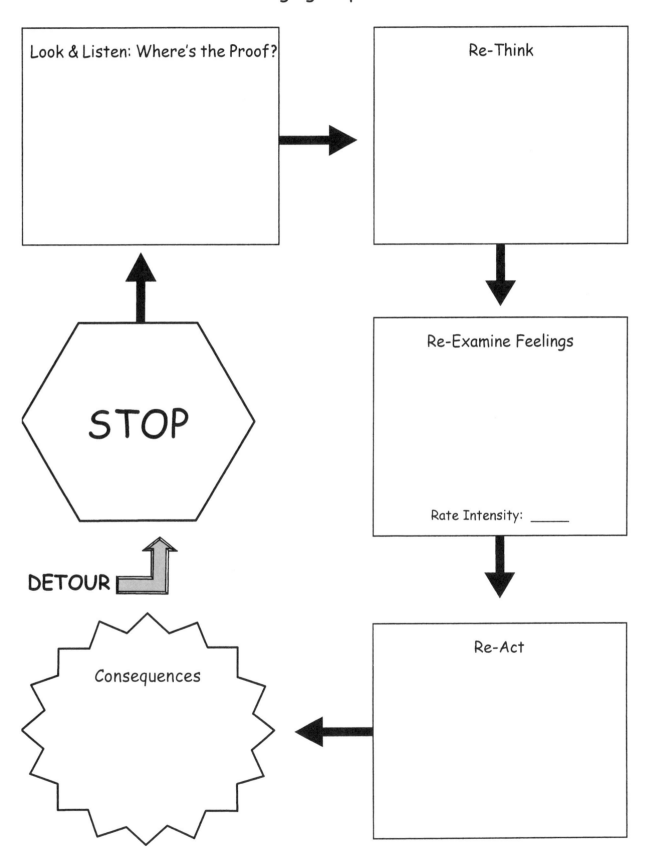

My Action Plan

The emotion(s) that has caused me difficulty is: _____

This emotion has stopped me from: _____

The small goal I would like to achieve is: _____

Goal(s)	Specific Steps to Take	Evaluate Results

My Action Plan

The emotion(s) that has caused me difficulty is: _____

This emotion has stopped me from: _____

The small goal I would like to achieve is: _____

Goal(s)	Specific Steps to Take	Evaluate Results

My Action Plan

The emotion(s) that has caused me difficulty is: _____

This emotion has stopped me from: _____

The small goal I would like to achieve is: _____

Goal(s)	Specific Steps to Take	Evaluate Results

My Action Plan

The emotion(s) that has caused me difficulty is: _____

This emotion has stopped me from: _____

The small goal I would like to achieve is: _____

Goal(s)	Specific Steps to Take	Evaluate Results

My Action Plan

The emotion(s) that has caused me difficulty is: _____

This emotion has stopped me from: _____

The small goal I would like to achieve is: _____

Goal(s)	Specific Steps to Take	Evaluate Results

My Action Plan

The emotion(s) that has caused me difficulty is: _____

This emotion has stopped me from: _____

The small goal I would like to achieve is: _____

Goal(s)	Specific Steps to Take	Evaluate Results

My Action Plan

The emotion(s) that has caused me difficulty is: _____

This emotion has stopped me from: _____

The small goal I would like to achieve is: _____

Goal(s)	Specific Steps to Take	Evaluate Results

My Action Plan

The emotion(s) that has caused me difficulty is: _____

This emotion has stopped me from: _____

The small goal I would like to achieve is: _____

Goal(s)	Specific Steps to Take	Evaluate Results

My Action Plan

The emotion(s) that has caused me difficulty is: _____

This emotion has stopped me from: _____

The small goal I would like to achieve is: _____

Goal(s)	Specific Steps to Take	Evaluate Results

Appendix C

Thought Records

Mood Management Thought Record

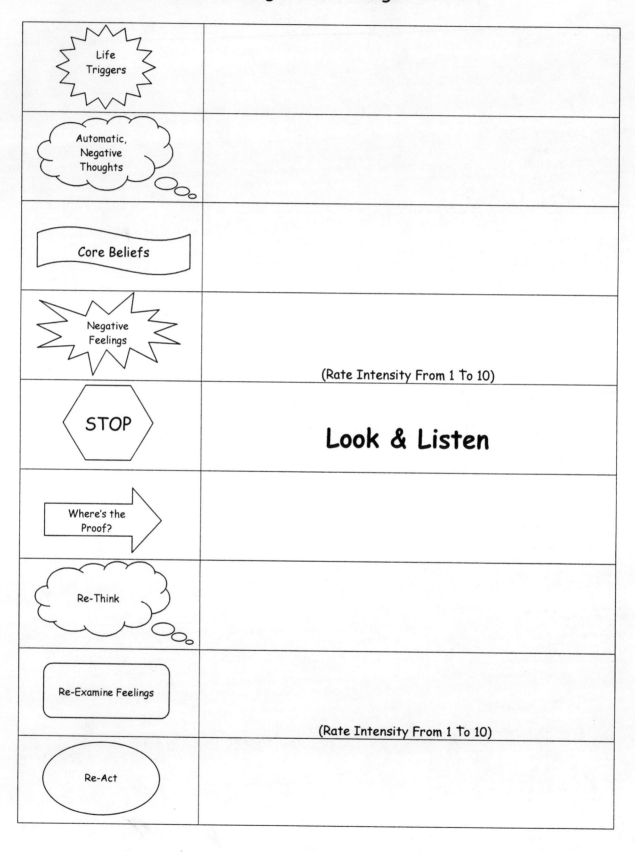

Life Triggers

Automatic, Negative Thoughts

Core Beliefs

Negative Feelings

(Rate Intensity From 1 To 10)

STOP

Look & Listen

Where's the Proof?

Re-Think

Re-Examine Feelings

(Rate Intensity From 1 To 10)

Re-Act

Mood Management Thought Record

Life Triggers	
Automatic, Negative Thoughts	
Core Beliefs	
Negative Feelings	(Rate Intensity From 1 to 10)
STOP	**Look & Listen**
Where's the Proof?	
Re-Think	
Re-Examine Feelings	(Rate Intensity From 1 to 10)
Re-Act	

Mood Management Thought Record

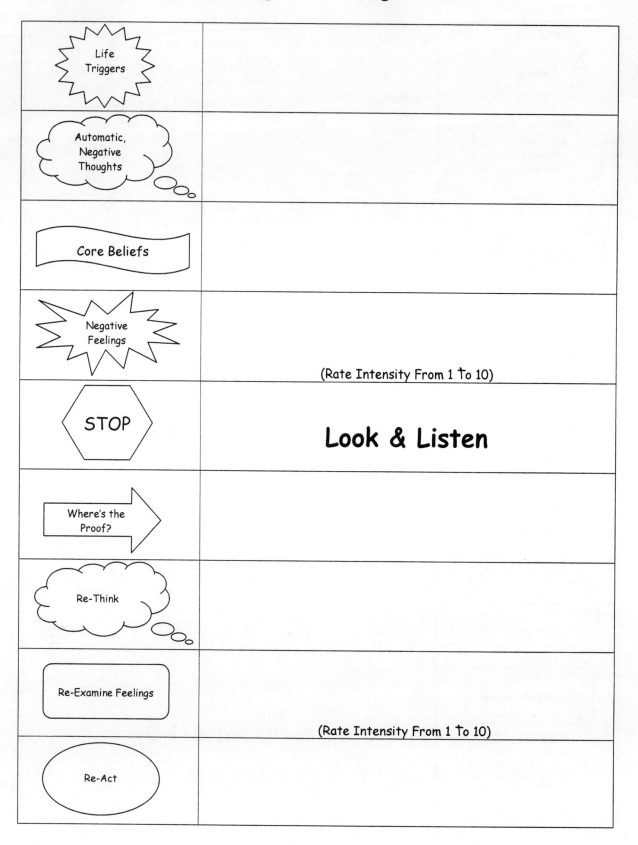

Life Triggers	
Automatic, Negative Thoughts	
Core Beliefs	
Negative Feelings	(Rate Intensity From 1 to 10)
STOP	**Look & Listen**
Where's the Proof?	
Re-Think	
Re-Examine Feelings	(Rate Intensity From 1 to 10)
Re-Act	

Mood Management Thought Record

Life Triggers	
Automatic, Negative Thoughts	
Core Beliefs	
Negative Feelings	(Rate Intensity From 1 to 10)
STOP	**Look & Listen**
Where's the Proof?	
Re-Think	
Re-Examine Feelings	(Rate Intensity From 1 to 10)
Re-Act	

Mood Management Thought Record

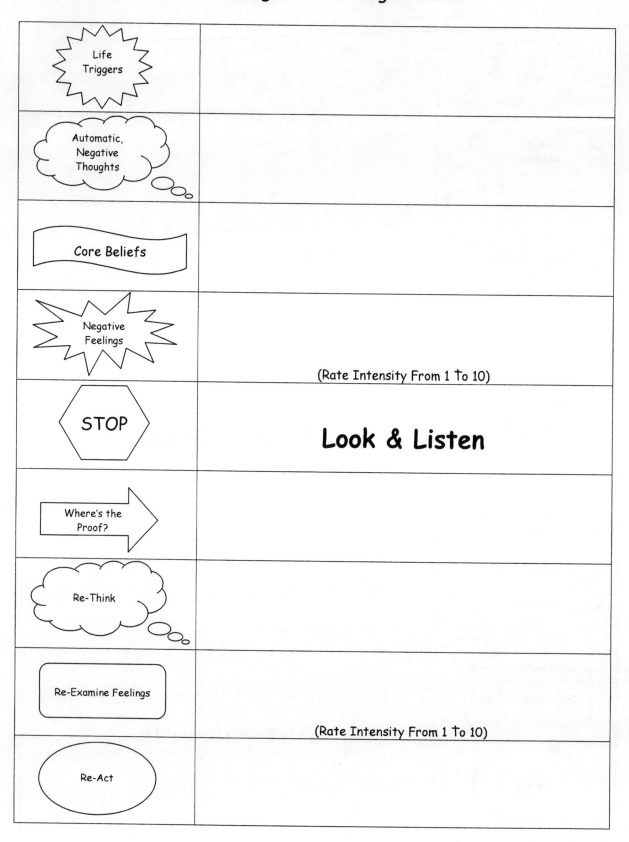

Life Triggers	
Automatic, Negative Thoughts	
Core Beliefs	
Negative Feelings	(Rate Intensity From 1 To 10)
STOP	**Look & Listen**
Where's the Proof?	
Re-Think	
Re-Examine Feelings	(Rate Intensity From 1 To 10)
Re-Act	

Mood Management Thought Record

Life Triggers	
Automatic, Negative Thoughts	
Core Beliefs	
Negative Feelings	(Rate Intensity From 1 to 10)
STOP	**Look & Listen**
Where's the Proof?	
Re-Think	
Re-Examine Feelings	(Rate Intensity From 1 to 10)
Re-Act	

Mood Management Thought Record

Life Triggers	
Automatic, Negative Thoughts	
Core Beliefs	
Negative Feelings	(Rate Intensity From 1 To 10)
STOP	**Look & Listen**
Where's the Proof?	
Re-Think	
Re-Examine Feelings	(Rate Intensity From 1 To 10)
Re-Act	

Mood Management Thought Record

Life Triggers	
Automatic, Negative Thoughts	
Core Beliefs	
Negative Feelings	(Rate Intensity From 1 to 10)
STOP	**Look & Listen**
Where's the Proof?	
Re-Think	
Re-Examine Feelings	(Rate Intensity From 1 to 10)
Re-Act	

Mood Management Thought Record

Life Triggers	
Automatic, Negative Thoughts	
Core Beliefs	
Negative Feelings	(Rate Intensity From 1 to 10)
STOP	**Look & Listen**
Where's the Proof?	
Re-Think	
Re-Examine Feelings	(Rate Intensity From 1 to 10)
Re-Act	

Appendix D

Transparency Masters

The 3 Yes Rule

☑ Do my goals involve changing myself rather than expecting others to change?

☑ Do my goals involve changing things that are within my control?

☑ Are my goals realistic?

Mood Management Skills Workbook - Unit 1 - Page 9

It's All About IMAGE

✓ **I** do care.

✓ **M**anaging problems effectively is a plus.

✓ **A**wareness is important.

✓ **G**o for it - it's worth the effort!

✓ **E**motions - we all have them and can learn to deal with them effectively.

Mood Management Skills Workbook - Unit 2 - Page 16

Making General Goals More Specific

✓ What would be different if I were approaching the goals I set in Unit 1?

✓ What changes would I see?

✓ What smaller steps are necessary in order to achieve my initial goal?

✓ What is giving me trouble now?

✓ How will I know when I'm doing better - what will happen?

Emotional Traffic Jams:
The Five Roads

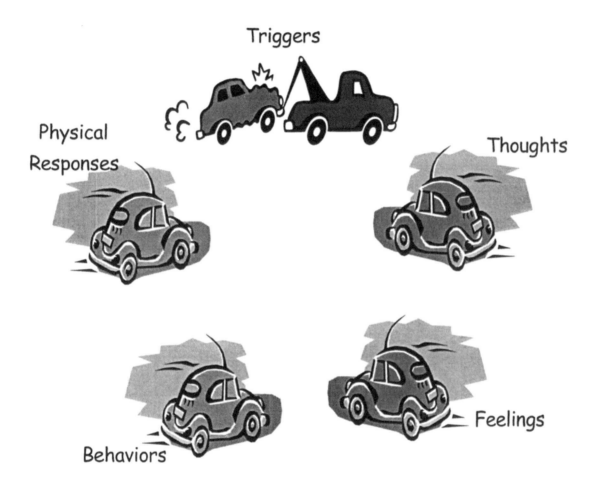

Wellness Mind - Emotional Mind

6.

It's the Thought That Counts!

Student "A"

Student "B"

Mood Management Skills Workbook - Unit 4 - Page 36

143

Thoughts, Feelings, & Behaviors

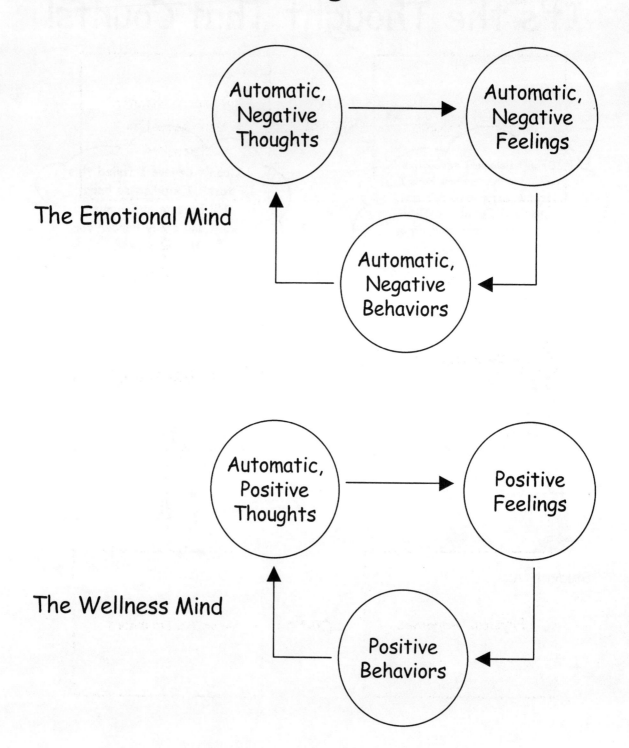

The Emotional Mind

The Wellness Mind

Negative Emotional Cycles

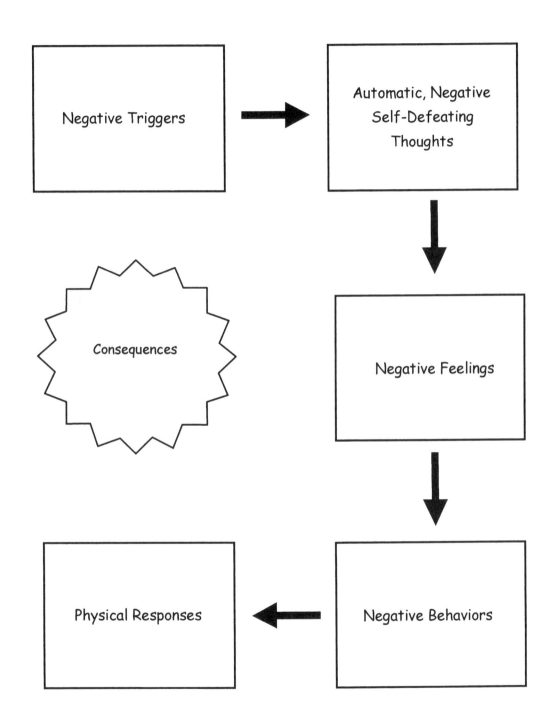

Negative Triggers

Automatic, Negative Self-Defeating Thoughts

Consequences

Negative Feelings

Physical Responses

Negative Behaviors

The Cycle of Depression

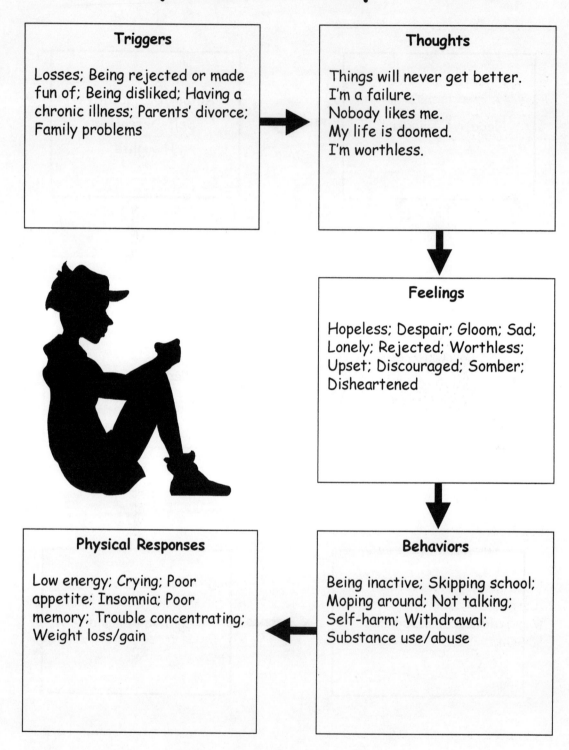

Triggers

Losses; Being rejected or made fun of; Being disliked; Having a chronic illness; Parents' divorce; Family problems

Thoughts

Things will never get better.
I'm a failure.
Nobody likes me.
My life is doomed.
I'm worthless.

Feelings

Hopeless; Despair; Gloom; Sad; Lonely; Rejected; Worthless; Upset; Discouraged; Somber; Disheartened

Physical Responses

Low energy; Crying; Poor appetite; Insomnia; Poor memory; Trouble concentrating; Weight loss/gain

Behaviors

Being inactive; Skipping school; Moping around; Not talking; Self-harm; Withdrawal; Substance use/abuse

Mood Management Skills Workbook - Unit 5 - Page 49

The Cycle of Anger

Triggers

Being rejected or made fun of;
Poverty; Emotional/physical
pain; Parents' divorce; Loss;
Chronic illness; Family problems

Thoughts

Everyone is out to get me.
Leave me alone.
Life is unfair.
I can't change.
I'll hurt you first.
I resent that.
I don't care.

Feelings

Irritable; Aggravated;
Hopeless; Rageful; Hurt;
Rejected; Hate;
Annoyed; Perturbed;
Exasperated; Riled up

Physical Responses

Tight muscles; Clenched fists;
Rapid heartbeat; Increased
blood pressure; Sweating;
Shaking; Trouble breathing

Behaviors

Frequent fighting; Substance
use/abuse; Self-harm; Arguing;
Blaming others; Being defensive;
Not doing schoolwork; Throwing
things

Mood Management Skills Workbook - Unit 5 - Page 50

The Cycle of Anxiety

Triggers

Disasters; Life changes (moving, death); Speaking in public; Automobile accident; Chronic illness; Physical/emotional pain

Thoughts

This is really scary.
I can't handle this.
Something bad will happen.
I'm helpless.
People always make fun of me.
I'll be too embarrassed.

Feelings

Afraid; Nervous; Irritable; Confused; Panicky; Tense; Apprehensive; Helpless; Embarrassed; Shaky

Physical Responses

Tight muscles; Rapid heartbeat; Increased blood pressure; Sweating; Flushed cheeks; Shortness of breath

Behaviors

Skipping school; Avoiding situations; Perfectionism; Substance use/abuse; Becoming dependent on others; Making up excuses

Mood Management Skills Workbook - Unit 5 - Page 51

The Cycle of Low Self-Esteem

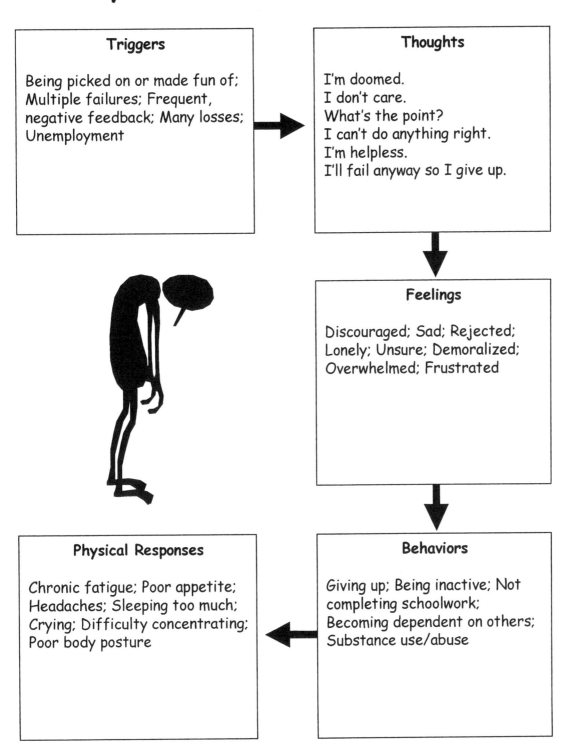

Triggers

Being picked on or made fun of;
Multiple failures; Frequent,
negative feedback; Many losses;
Unemployment

Thoughts

I'm doomed.
I don't care.
What's the point?
I can't do anything right.
I'm helpless.
I'll fail anyway so I give up.

Feelings

Discouraged; Sad; Rejected;
Lonely; Unsure; Demoralized;
Overwhelmed; Frustrated

Physical Responses

Chronic fatigue; Poor appetite;
Headaches; Sleeping too much;
Crying; Difficulty concentrating;
Poor body posture

Behaviors

Giving up; Being inactive; Not
completing schoolwork;
Becoming dependent on others;
Substance use/abuse

Mood Management Skills Workbook - Unit 5 - Page 52

Characteristics of the Emotional Mind

✓ It generates automatic, self-defeating thoughts.

✓ It wants you to believe negative things about yourself, your future, and your world.

✓ It's fast and furious. 0 to 60 in 8.2 seconds.

✓ It likes to trick you.

✓ It uses key words such as "never," "should," "always," "if/then," and "everything."

✓ It keeps you stuck in your negative emotional cycle.

✓ It often gives you the same interpretation of different triggers that, over time, causes core beliefs to develop.

Challenging: Be Your Own Mood Police

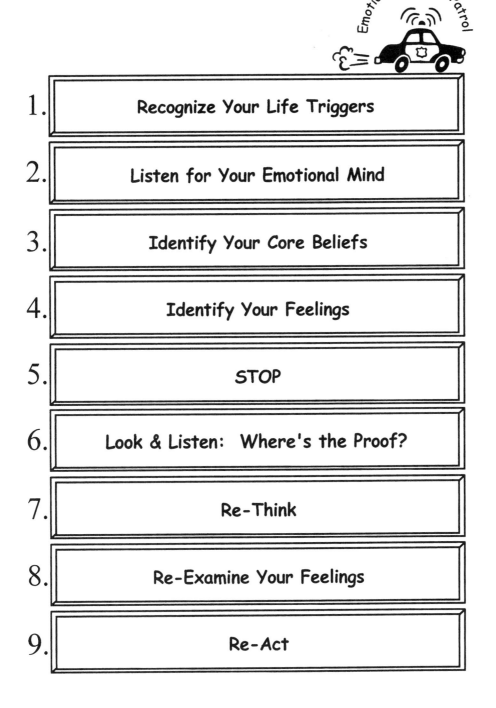

Emotional Wellness Patrol

1. Recognize Your Life Triggers

2. Listen for Your Emotional Mind

3. Identify Your Core Beliefs

4. Identify Your Feelings

5. STOP

6. Look & Listen: Where's the Proof?

7. Re-Think

8. Re-Examine Your Feelings

9. Re-Act

Mood Management Skills Workbook - Unit 6 - Page 72

Challenging: Take a Detour From Your Emotional Mind

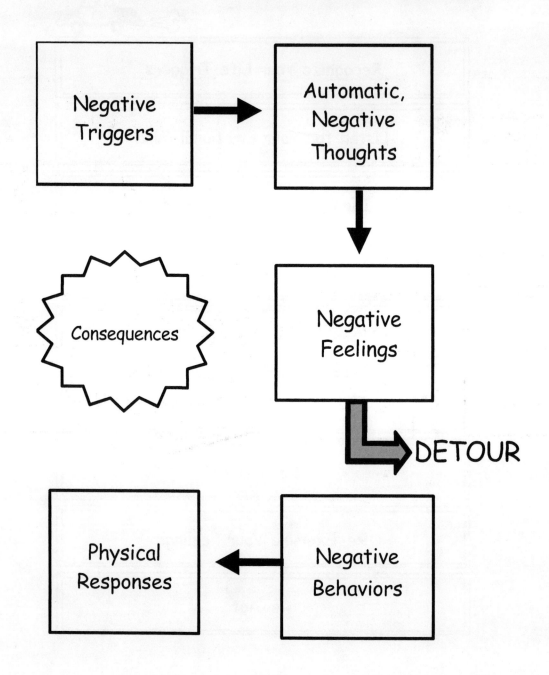

Challenging Map: The Route to the Wellness Mind

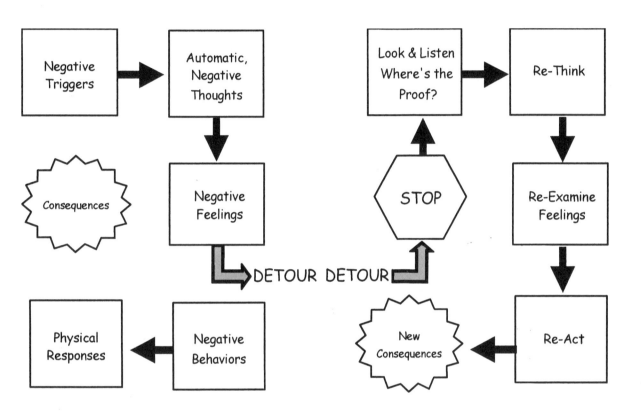

The Thought Record

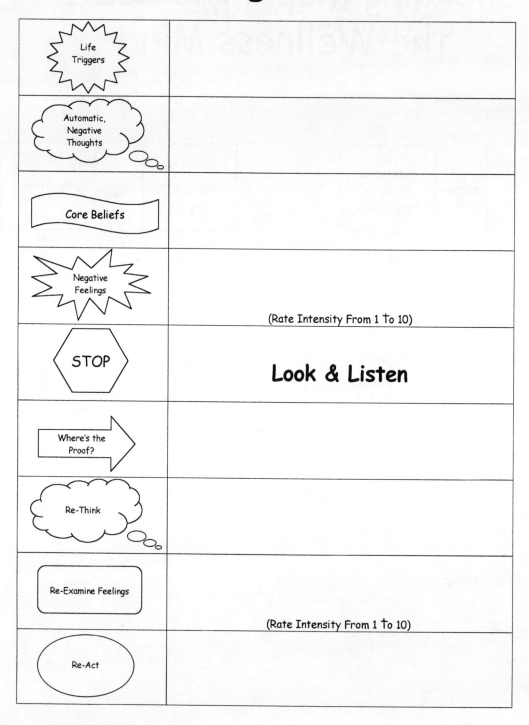

Life Triggers	
Automatic, Negative Thoughts	
Core Beliefs	
Negative Feelings	(Rate Intensity From 1 to 10)
STOP	**Look & Listen**
Where's the Proof?	
Re-Think	
Re-Examine Feelings	(Rate Intensity From 1 to 10)
Re-Act	

Mood Management Skills Workbook - Unit 7 - Page 90

My Action Plan...

The emotion(s) that has caused me difficulty is: <u>anger</u>

This emotion has stopped me from: <u>getting along with others</u>

The small goal I would like to achieve is: <u>getting along with one person</u>

Goal(s)	Specific Steps	Evaluate Results
Get along better with my boyfriend/girlfriend.	1. Listen carefully to what he/she says. 2. Buy him/her a card just because. 3. When I feel angry, count to 10 before saying anything. 4. When I feel angry, tell him/her that I need a time-out. 5. When I feel angry, fill out a Challenging Map.	1. No problems. 2. He/she was surprised. 3. I had to count to 30.

Mood Management Skills Workbook - Unit 7 - Page 93

About the Author

Carol A. Langelier, PhD, is a licensed psychologist and member of the adjunct faculty at Rivier College, where she teaches courses in the Graduate Counseling and School Psychology programs. She is employed as a school psychologist in New Hampshire. Her experience working with adolescents is extensive. She has been employed in clinical inpatient and outpatient settings as well as schools during her career as a teacher, counselor, and psychologist. She is a member of the American Psychological Association, the American Counseling Association, and the National Association of School Psychologists. Her interests include multicultural counseling. She has authored and coauthored articles in this area and has presented at national conferences on multicultural identity development. She also works as a diversity trainer, providing workshops to schools on the cultural context of education.